The Interactions of Human Mobility and Farming Systems on Biodiversity and Soil Quality in the Western Highlands of Cameroon

I0110132

Christopher Mubeteneh Tankou

Langaa Research & Publishing CIG
Mankon, Bamenda

Publisher:
Langaa RPCIG
Langaa Research & Publishing Common Initiative Group
P.O. Box 902 Mankon
Bamenda
North West Region
Cameroon
Langaagrp@gmail.com
www.langaa-rpcig.net

Distributed in and outside N. America by African Books Collective
orders@africanbookscollective.com
www.africanbookcollective.com

ISBN: 9956-791-89-X

DISCLAIMER
All views expressed in this publication are those of the author and do not necessarily reflect the views of Langaa RPCIG.

Table of Contents

Tables and Figures

Tables

Figures

1 General Introduction

1.1 Introduction

Human mobility takes a wide variety of forms from permanent migration between regions to diurnal movements such as commuting to work (Parnwell, 1993). These movements generate flux in population numbers and carve out pathways in time and space. Capturing these movements is critical to understanding the use of space and place as mobility almost inevitably affects people's differential access to social, economic and human resources (De Haas, 2009). Human beings therefore, are constantly on the move, transgressing social and spatial boundaries to expand their capabilities and entitlements in order to improve their welfare (Sen, 1999). Movements in relation to rural land use may have a widely diverging impact on the biodiversity and farming systems in the rural milieu, depending on the type and causes of the movement and the interactions with farming systems (Quinn, 2006; Stark & Taylor, 1989; Taylor, 1999). While the consequence of migration on national development in "sending" countries is generally the loss of the highly skilled labour force ('brains'), the loss of an agricultural local workforce ('brawn'), affects the standard of living and farming systems in rural areas of the "sending" country. The departure of young, able-bodied men and women from rural areas (Lewis, 1986) has been typically blamed for causing a shortage of agricultural and other labour, depriving areas of their work force and thus affecting farming systems (Lipton, 1980; Rubenstein, 1992; Taylor, 1984), which can disrupt economic production and lead to a decline of productivity in the agricultural and other traditional sectors (De Haas, 1998). The causes and course of intra-rural and rural-urban mobility reflect the effects of contrived economic policies that have systematically marginalised the rural sector. Governments' neglect of this sector fuels unemployment, low productivity, poverty and rural exodus. Studying human mobility and land use as discrete subjects of inquiry precludes a comprehensive understanding of the process by which rural households attempt to maximize resource access. If we are interested in understanding the variables that influence land use change, we must inquire not only how people are managing land, but why they come to be there in the first place and hence the genesis of certain patterns of mobility. Studies have demonstrated that people from drier regions are more likely to move temporarily and permanently to other rural areas (rural-rural movement), compared to people from wetter areas and that long-term movement seems to be less related to environmental conditions than short-term moves (Henry *et al.*, 2004). In general, scarcity of land influences the rural-rural mobility of farmers from land-deficit to land-surplus areas (Swindell, 1984). Fragmented, unproductive landholdings, climate change and poor incomes compel farmers to seek more fertile land, land with special advantages (such as land close to streams that can be used for irrigation during the off-season), wage labour or non-farm activities (Pike & Rimmington, 1965; Pike, 1968).

Rural-urban and rural-rural mobility affect both rural and urban change, and influence resource use and management (Potts, 2006). Urban-rural and intra-rural linkages include flows of people, of goods, of money and of information, as well as other social transactions that are central to social, cultural and economic transformation. Within the economic sphere, many urban enterprises rely on demand from rural consumers, while access to urban markets is often crucial for agricultural producers (Kydd & Christiansen,1982; Kalipeni, 1992). In addition, a large number of urban-based and rural-based households rely on the combination of agricultural and non-agricultural income sources for their livelihoods. This may involve straddling the rural-urban divide from a spatial point of view involving movement between urban and rural areas, or from a sectoral point of view engaging in agriculture in urban centres or in non-farm activities in rural areas (Tacoli, 2002). A traditional movement of population in Africa which is worth mentioning concerns the sahelian population, whose mobility is dictated by vagaries of climate and poverty (De Bruijn & van Dijk, 2003). Their mobility which is a typically rural-rural type, is southbound in search of favourable grazing grounds for their cattle due to the limited and irregular rainfall pattern of the zone and also due to the uneven economic development of the sahelian and coastal countries in west Africa which favours the movement of Fulbes to the coast in search of a better livelihood (*ibid*). However there is a dearth of current migration data in Africa, a continent shaped by migration over a number of centuries. As a result, many statements about African mobility are based more on supposition rather than on empirical evidence. This combination of the lack of empirical data and the use of inappropriate conceptual frameworks has contributed to distorted and highly simplistic views on the nature of African mobility. More empirical research on African mobility is absolutely essential in order to achieve an improved understanding of the subject.

Human land-use and its influence on land cover, is a major factor in the distribution and functioning of ecosystems, and thus in the delivery of ecosystem services (Costanza *et al.*, 1997). Activities such as agriculture, forestry, transport, manufacturing and housing, alter the natural state and ecosystem functions of land, as they involve land-cover conversion or land-use intensification. Thus agricultural intensification affects biodiversity, water resources and soil quality, and contributes to greenhouse gas emissions (EEA, 2006; Ramankutty, 2010). The tropics have the highest biodiversity and tropical forest areas have the largest volume of biomass. However, tropical natural biodiversity, ecosystem functioning and ecosystem services are directly and indirectly threatened by population pressure and a broad variety of human activities, especially land use and mobility, both within and outside of the rural areas (Costanza et al., 1997). It has been estimated that one-fifth of extant vertebrate species are classified as threatened, ranging from 13% of birds to 41% of amphibians (Hoffmann et al., 2010). The per cent increase of Africa's population between 2005 and 2050 of 117 is considered to be the highest compared to other regions during the same length of time (UN, 2007). Studies of land-cover changes in West-Africa show that agricultural expansion is the most dominant trajectory of land-cover change which involves loss of savannah and forest (Wood et al., 2004; Braimoh & Vlek, 2005). Several concepts have been proposed to describe the relationship, functioning and feedback of land-cover (environment), land-use (economy) and socio-cultural conversions (e.g. deforestation) and modifications (e.g. changing land use management such as fertilizer use and irrigation practices) of land-use. These relationships have significance for the functioning of the earth's ecosystems through their impact on biogeochemical cycles (Turner et al., 1994) and important consequences for food security (Brown, 1995).

Cameroon is an agricultural economy and the rural sector, which accounts for 30% of GDP, plays a significant role in her economy (Anonymous, 2008). Despite its importance in the economy of the nation, the rural sector faces many problems that account for declining productivity and biodiversity (Anonymous, 2008) orchestrated by population pressure, mobility and improper land use. From the late 1970s through to 1985 Cameroon was a booming economy. Beginning in 1985 it experienced a devastating economic crisis, a structural adjustment program in 1989 and, in January 1994, a drastic devaluation of its currency (CFA) (Sunderlin & Pokam, 1998). The CFA Franc was devaluated in 1994 by 50% to restore the competitiveness of exports which, coupled with increased world prices for cocoa and coffee, led to a mild resurgence in the profitability of these commodities (Sunderlin & Pokam, 1998). In 1989, the purchase prices of cocoa and coffee (the leading agricultural export earners) were cut in half by the government, greatly adding to the impoverishment of farmers. These economic transformations profoundly influenced population movements and land-use (ibid).

The functioning of any individual farming system is strongly influenced by the external rural environment, policies, institutions, markets and information linkages (Dixon, 2001). Differences in farming systems and land use can alter nutrient input and output fluxes, in soil and vegetation. This can change soil fertility, which in turn affects biomass production and human decisions on land management (Priess, 2001). Nutrient-efficient farming is characterized by the minimization of nutrient losses to the environment while ensuring the necessary nutrient supply to crops and livestock. The tropics and particularly the humid savannah agro-ecological zone in West and Central Africa are generally nutrient-poor ecosystems with potentially low productivity due to erratic rainfall which favour soil degradation. The Western Highlands of Cameroon (WHC) which covers the North-West and West regions however consist of rich volcanic soils coupled with high rural population densities of 48.4 and 59.8 persons per square kilometre respectively. This has contributed to the development of varying patterns of human mobility and land use changes, with major impacts on farming systems, biodiversity and soil nutrient balance in this agro-ecological zone. In this area, farmlands are intensively used due to population pressure. This implies, in spite of higher levels of soil fertility, high inputs of mineral fertilizers and pesticides. In general, production norms for these crops are virtually non-existent and most farmers apply inputs based on their financial capabilities. Due to the shortage of fertile land and water sources for irrigation, human mobility and land-use interactions have reshaped the farming systems, soil nutrient dynamics and biodiversity in the WHC. Agricultural intensification in this zone has led to the switch from 'traditional' to 'modern' farming systems, characterized by the increased use of mineral fertilizers and pesticides. Many farmers in this zone have limited access to inputs but are forced by circumstances to drastically reduce the complexity of their agro-ecosystems in an attempt to intensify production; even though the maintenance of crop diversity is widely accepted as a means of buffering farmers against short-term crop failures (Willey, 1979). The natural biodiversity and the agro biodiversity of the WHC have changed drastically due to forest clearance for agriculture, both in composition and structure, but also more recently due to the replacement of the traditional shifting cultivation with multiple species, by intensive short fallow cultivation systems characterized by sole cropping. The traditionally preserved 'sacred groves' of the area have conserved some of the endemic plant diversity (Khan, 1997),and these sacred groves still have an important function in terms of traditional medicines and the provision of fruits and forest products (Khan, 1997). A major feature of population pressure and consequently human mobility in the WHC is the intensification of land-use for food and feed production. One of the great challenges is to gain a sound scientific understanding of human mobility and land interactions and feedback processes in the complex and highly diverse tropical ecosystems in general, and the humid savannah agro-ecological zone of Cameroon in particular, and its impact on local natural biodiversity and agro biodiversity. Improved knowledge in this field through education must form the basis of any plan to optimize farming systems and land-use planning and halt the accelerating biodiversity loss in the tropics in general and Cameroon in particular, and to improve soil nutrient balance and the sustainability of the farming systems. No studies have been carried out which attempt to understand the relationship between human mobility and land-use interactions in the WHC, considered to be the breadbasket of the Central African sub region, and which has been forced to adjust after severe pressure imposed by socio-politico-economic factors.

1.2 Objectives

The main objective of this study is to determine the interactions between human mobility and farming systems, and the impact on local plant diversity and soil nutrient balance to prevent land degradation and improve sustainability.
- Different land-use practices and abiotic factors.
- Analyse crop and farm-level nutrient balances.

The research questions have been defined as follows:
- What are the forces driving human mobility and their contribution to the development of the different categories of human mobility in the study area?
- What are the levels of sustainability of farming systems and the relationship between sustainability and the different forces driving farming systems in this zone?
- What are the different types of plant diversity of agro-ecosystems and how are they influenced by abiotic factors?
- What impact does the the modification of the farming system have, on soil quality at the crop and farm levels of the study area?

1.3 Conceptual framework

The conceptual framework (Figure 1.1) describes the paradigm linking the elements of the framework. The IPAT identity (Waggoner & Ausubel, 2002) which offers a comprehensive identification of "driving forces" can be used to understand this framework. This relationship presented as population = environmental impact, incorporates both the affluence (A) of the population (P) in question and the technology (T) with which people affect their impact (I). The IPAT equation (I = PAT) inocorporates the combined interaction rather than independent effects in determining environmental change. Farmers in the highly dense WHC have had to adjust to an unsustainable drop in coffee prices on the international market by adopting vegetable cash crop cultivation. The study relies on the hypotheses that the main factors leading to varying household responses, are increasing population and stress imposed by economic factors. The responses to these forces include the expansion of cultivated land and residential areas. Also, the economic hardship that resulted from the drop in market prices of the only cash crop of the study area, coffee, drastically reduced income generation in the rural areas. The production system that reigned when coffee was the main cash crop was long-fallow shifting cultivation or swidden system. Global expansion of cultivated land (conversion) accelerated, along with the intensification of the use of land already cultivated (modification). Neoclassical economics typically accounts for the role of population change through its influence on demand as manifested through the market. When the market signals change, the land-use also changes (Meyer & Turner, 1992).

Swidden cultivation is a natural resource practice strategy that involves the rotation of fields rather than crops and relies on the use of fallow to sustain the production of food crops (Nielsen, 2006). The fallows are cleared by means of slashing and burning, the land is cropped for a short period of time and then left untended while the natural vegetation regenerates. Regardless of a lack of substantiating data, swidden systems have frequently been deemed to be environmentally destructive causing deforestation, soil degradation in terms of erosion and negative nutrient balances and contributing to CO_2 emissions (Brady, 1996; Devendra & Thomas, 2002; Harwood, 1996). This perception is however, increasingly challenged as numerous studies have shown that swidden cultivation in many situations can be a rational economic and environmental choice for resource poor farmers (especially with regard to labour), and that swidden cultivation besides being a production system, provides a range of ecosystem services in terms of hydrology, biodiversity and carbon storage in soil and vegetation (Fox, 2000; Kleinman, 1995, 1996; Nielsen, 2006).

Under the swidden system, there was virtually no demand for off-farm (farm inputs obtained outside the farm such as fertilizers and pesticides bought in the markets and transported to the farm) inputs owing to the natural nutrient regeneration processes and the elimination of pests through long term absence of host plants. Farmers depended on their previous harvests for planting materials which made the system nearly totally dependent on on-farm (farm inputs generated from the farm such as plant nutrients obtained through fallowing and pest control through the elimination of pests during the fallow period) inputs. One of the responses to mitigate this economic crisis was the search for high-income generating cash crops which could be considered as a micro-level factor responsible for mobility (De Jong & Gardner, 1981). Cool season vegetable crops, highly solicited by national and international markets fulfilled the requirements and became the new important cash crops of the WHC.

However, in order to improve livelihoods, another response was the emergence of different types of human mobility. Some household members in WHC migrated to urban centres while some others remained in the rural areas but improved their livelihoods through commuting to favourable cash crop production areas. Hence rural-rural mobility led to a massive occupation of the high altitude zones of the WHC for cool-season crop production and the construction of rural road infrastructure, practical mostly during the dry season and renovated only during periods of political campaigns. Owing to the scarcity of land favourable to these new cash crops, the long fallow production system became technically unfeasible and was replaced to a great extent by intensive land use systems highly dependent on the use of agro-chemicals for the production of the vegetable cash crops. Hence, the interaction between human mobility and farming systems provoked significant modifications, leading to an impact on the farming system, significantly reflected in the change in natural and agro-biodiversity and the soil quality of the zone. In view of the significance of agriculture in the WHC, population dynamics, especially human mobility, could be considered in an ecological context. Agriculture has been the lifeline of the people and for many decades a delicate balance was maintained between the extensive, long-fallow agricultural production system and the fragile environment. But the current prevailing human mobility trends attest to an acute and potentially disastrous imbalance between land-use and natural ecosystems, since rapid population growth increased the demand for land and resulted in intensive land use which eliminated the traditional on-farm dependent system.

1.4 Theoretical framework

One school of thought is the Clifford Geertz's concept of "agricultural involution" (Geertz, 1963) based on the internal complexity within static socio-economic forms in historical Indonesia. Innovative intensification is referred to as increasing the amount of time land is under cultivation in addition to adopting new crops or new techniques, while non-innovative intensification involves increasing cropping intensity while maintaining the same cultivar (Laney, 2002). Non-innovative intensification encompasses increasing inputs without a concurrent techno-managerial shift, which in the extreme leads to a levelling or possibly a decrease in output per unit of input, known as stagnation or involution, respectively (Geertz, 1963). Non-innovative intensification often involves an increase in the crop cover over the course of a given year through a reduction in the length of fallow, while innovative intensification may involve this and/or a shift to a new cultivar, the application of chemical fertilizers and pesticides, the use of petroleum-fuelled machinery, and/or significant alteration in local hydrology. In their analysis of the change in cropping intensities, Keys and McConnell (2005) found that most non-innovative intensification occurred in Africa. African small scale farmers utilize family labour to increase labour input despite decreased marginal productivity. This describes a process of involution and does not clash, in the short-term, with strategies to increase production per capita (Geertz 1963). But if the process continues and more labour is applied on the same parcel of land, the effect will sooner or later be a decreased production per capita and thus an emerging agrarian crisis.

The demographic pressure can have a very severe impact on natural ecosystem services through associated human mobility and changing farming systems. It has become increasingly recognised that a major factor in both commuting and migration is on-going environmental degradation induced by population pressure. Bilsborrow (1979, 1992) suggested a framework that integrated the Malthusian theory that a growing population demographically responded to resource pressure by fertility reduction or out-migration that ultimately reduced resource demands and the Boserupian theory that a population economically responded to resource pressure through changes in agricultural technology that ultimately increased supply. To do this, Bilsborrow (1992) draws on Kingsley Davis's concept of "multiphasic response". Although the response of a growing population could be multiphasic by involving any combination of the demographic, economic and demographic-economic changes considered, Bilsborrow (1992) noted that taking up one response meant the others were less likely to be taken up, because pressure would then be reduced, and the stimulus mollified. According to Von Thunen's classic "concentric zones" model, landscape change is seen as a response to changes in economic land rents associated with increasing transportation costs from the farm to a central market place (Bryant, 1982).

Mukherji Shekhar (2001) proposed a theoretical system perspective approach, and argued that need-attribute systems of the people, utility offerings of the places and different mobility behaviour that arise to satisfy needs, are interdependent parts of a system called the mobility field, and postulated that any natural or induced change in any part of the system would generate corresponding changes in other parts. The need systems of the individuals are regarded as the causal forces acting behind their movement behaviour, and consequently, it is suggested that if it is possible to induce desirable changes in the need-stress attribute structure of the people, then it can effect changes in people's spatial behaviour, and vice-versa. It is also assumed that by inducing change in the spatial arrangement and the utilities of the places, it is possible to induce changes in the behaviour of the people and in their attribute structure. Hence, mobility can be planned to act as an agent of socio-economic change.

Ecosystem services are generated by ecosystem functions which in turn are underpinned by biophysical structures and processes referred to as "supporting services" by the Millennium Ecosystem Assessment (2005). Ecosystem functions are thus intermediate between ecosystem processes and services and can be defined as the "capacity of ecosystems to provide goods and services that satisfy human needs, directly and indirectly" (De Groot, 1992). The dimensions of biodiversity and its relationship to human well-being have been extensively addressed by Levin (2000), including both the services that biodiversity supports and the evolutionary genesis of biodiversity, together with the ecological processes underlying patterns and trends. Ecosystems can be evaluated through the use of indices to measure biodiversity or through the measurement of nutrient flux. Genetic diversity of crops increases production and decreases susceptibility to pests and climate variation (Ewel, 1986; Altieri, 1990; Zhu, 2000). In low-input systems especially, locally adapted varieties often produce higher yields or are more resistant to pests than varieties bred for high performance under optimal conditions (Joshi, 2001).

Influenced by human management, ecosystem processes within agricultural systems can provide services that support the provisioning services, such as pollination, pest control, genetic diversity for future agricultural use, soil retention, regulation of soil fertility and nutrient cycling. The potential of an agricultural system to provide such services depends on the degree of sustainable management (Power, 2010).

Management practices also influence the potential for the impairment or 'disservice' of agriculture, such as loss of habitat for conserving biodiversity, nutrient runoff, sedimentation of waterways, and pesticide poisoning of humans and non-target species (Zhang, 2007). Since agricultural practices can harm biodiversity through multiple pathways, agriculture can be considered as harmful to conservation; however, appropriate management can ameliorate many of the negative impacts of agriculture, while largely maintaining provisioning services (Power, 2010). Agricultural regulating services include, flood control, water quality control, carbon storage and climate regulation through greenhouse gas emissions, disease regulation, and waste treatment (e.g. nutrients, pesticides). Cultural services include scenic beauty, education, recreation and tourism, as well as traditional use. Agricultural places (such as the sacred groves) or products are often used in traditional rituals and customs that bond human communities. Conservation of biodiversity may also be considered a cultural ecosystem service influenced by agriculture, since most cultures recognize the appreciation of nature as an explicit human value. In return, biodiversity can contribute a variety of supporting services to agroecosystems and surrounding ecosystems (Daily, 1997).

Around the world, there are great variations in the structure and function of agricultural ecosystems because they were designed by diverse cultures and under diverse socioeconomic conditions in diverse climatic regions. Functioning agroecosystems include, among others, annual crop monocultures, temperate perennial orchards, grazing systems, arid-land pastoral systems, tropical shifting cultivation systems, smallholder mixed cropping systems, paddy rice systems, tropical plantations (e.g. oil palm, coffee, and cacao), agroforestry systems and species-rich home gardens. This variety of agricultural systems results in a highly variable assortment and quantity of ecosystem services. Just as the provisioning services and products that derive from these agroecosystems vary, the support services, regulating services and cultural services also differ, resulting in extreme variations in the value these services provide, inside and outside the agroecosystem. In maximizing the value of provisioning services, agricultural activities are likely to modify or diminish the ecological services provided by unmanaged terrestrial ecosystems, but appropriate management of key processes may improve the ability of agroecosystems to provide a broad range of ecosystem services (Powel, 2010).

Globally, most landscapes have been modified by agricultural activities and most natural, unmanaged ecosystems are potential agricultural lands. The conversion of undisturbed natural ecosystems to agricultural use can have a strong impact on the system's ability to produce important ecosystem services, but many agricultural systems can also be important sources of services. Indeed, agricultural land use can be considered an intermediate stage in a human impact continuum between wilderness and urban ecosystems (Swinton, 2007). Just as conversion from natural ecosystems to agriculture can reduce the flow of certain ecosystem services, the intensification of agriculture (Matson, 1997) or the conversion of agro ecosystems to urban or suburban development can further degrade the provision ofbeneficial services.

In the WHC, the forces of demographic and economic pressures on the rural population have resulted in a variety of responses.

1.5   Structure of thesis

Chapter 2 deals with the response component of Figure 1.1 and provides case study findings through a structured questionnaire on human mobility in WHC, with an emphasis on the types of human mobility and what changes them, as well as their impact on land-use and occupational diversification in the research site. Chapter 3 constitutes the main impact in Figure 1.1 that examines the impact of the modification of the farming system in the case study locations with a focus on inter-household differences with respect to access of assets, especially farming inputs and their impact on sustainability. In addition, the major factors that influence the farming systems are identified and used to define the sustainability of the research site. The later two chapters examine in detail the main impact component of Figure 1.1. Chapter 4 describes in detail, patterns of agro biodiversity as influenced by types of land use and abiotic factors in the study areas. Chapter 5 focusses on exploring changing trends in soil quality at the crop and farm levels, imposed by the modification of the farming system. The final chapter summarises the key issues related to the main findings of the studies.The factors which driver socio-ecological change included the impact of population pressure on the varying types of human mobility and their effects (Box 1.1).

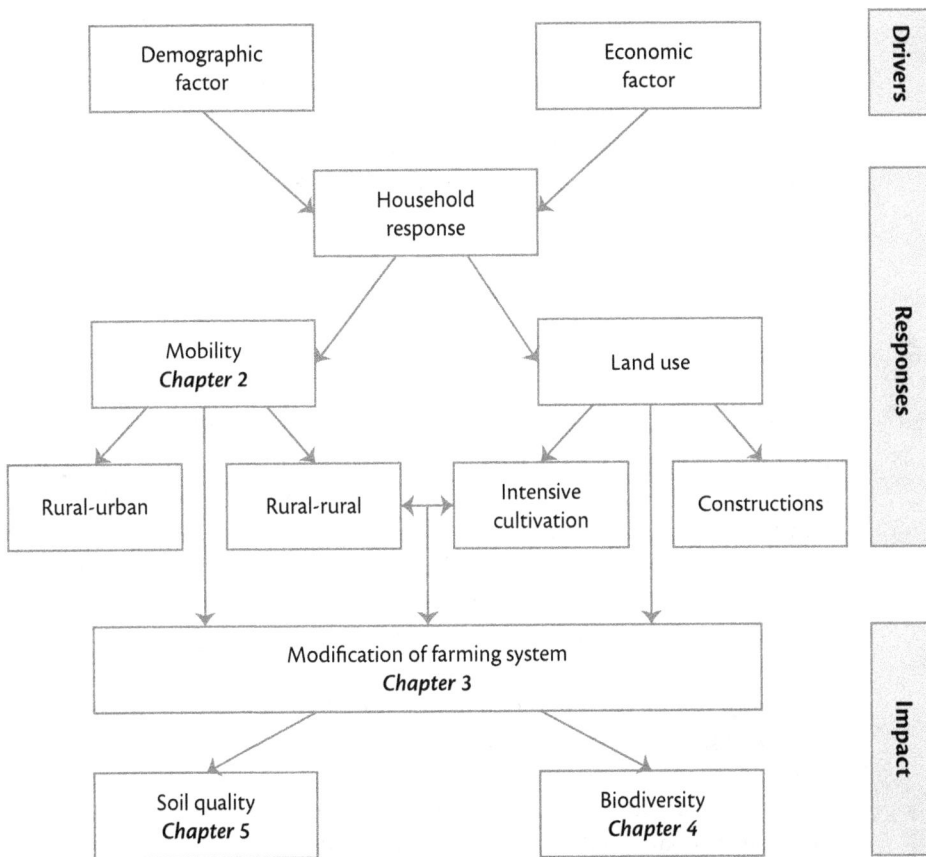

Figure 1.1
Simplified picture of the interaction of mobility and land-use in the WHC.

BOX 1.1
Socio-ecological sustainability
This research seeks to emphasize the environmental perspective with more attention focussed on how human mobility and farming systems address soil health, natural and agro biodiversity and agro-ecological processes that employ closed loop
cycles within the farm where possible, to limit inputs and reduce waste. Another obvious component is reducing environmental impacts, enhancing our natural
capital and building on environmental services on which our food system depends.
Wells (2011) posited that agriculture is an integral nexus of society and ecology
over time, a coevolution of culture and nature, humans and landscape. An attempt
to broach sustainable agriculture therefore, demands attention to its socio-ecological nature (Wittman, 2009). The social indicators of sustainable agriculture usually
underpin social, economic and ecological systems interacting at multiple temporal, spatial and organizational levels (Bacon et al. 2012). The object of this research
is an attempt at understanding this, through the analyses of the socio-ecological
impacts provoked by demographic pressure and cash crop crisis, in addition to the
existing types of human mobility on sustainability indicators in the study area. One
of the striking outcomes of urban-rural mobility is the genesis of social or cultural
mobility (Chapter 2).

1.6   The study site

The study was conducted in two sub-divisions in the WHC namely Fongo
Tongo and Nkong-ni all of which are found in the WHC, one of the major
agricultural zones of Cameroon. These two sub-divisions contain all the main
features found in the humid savannah zone of Cameroon. The climate of the
zone is tropical, with a mono-modal rainfall distribution. The growing season
is between mid March and mid November and the dry season is between mid
November and mid March. The annual rainfall is between 1000 and 2000mm,
annual maximum temperature is 22°C and annual minimum temperature is
17°C (Kay, 1985). The soils are ferralitic with high moisture retention capacities (Fotsing, 1992).The altitudinal range of the research sites is between 1400
m and 2000 m and more, above sea level. Cool season vegetable crop production is dominant at higher altitudes. Agricultural activities are also very
common in inland valley swamps and steep slopes. The population density
is approximated to 59.8 inhabitants per km². The natural vegetation is dominated by grass with shrubs and trees. Many households plant their land with
eucalyptus trees (most often for border demarcation) commonly using the
wood as timber and firewood. Production systems vary with the altitude. At
the lower altitudes, intercropping is predominant and crops (both warm and
cool season) grown are principally for home consumption while at higher alti-

tudes, sole cropping is more practised and the crops (cool season) grown, are more for the market. Generally people tend to move from the lower altitudes towards the higher altitudes in search of farmland for commercial production and/or labour jobs in commercial farms. Detailed methodologies for the different components of the thesis are elaborated in the various chapters.

References

Adams, M.L., Norvell, W.A., Philpot, W.D. & Peverly, J.H. (2000) Spectral detection of micronutrient deficiency in 'Bragg' soybean. *Agron. J.* 92(2): 261-268.

Adepoju, A.(2001) Population and Sustainable Development in Africa in the 21st century: Challenges and Prospects. HRDC *African Policy Research Series No. 1.* Lagos: Concept Publications.

Adepoju, A. & Hammar, T. (eds.) (1996) *International Migration in and from Africa: Dimension, Challenges and Prospects.* Dakar: PHRDA/Stockholm: CEIFO.

Altieri, M.A. (1990) Agroecology and rural development in Latin America. In: Altieri, M.A. & Hecht, S.B. (eds.) *Agroecology and Small Farm Development.* Florida: CRC Press, pp. 113-118.

Anonymous (1993) Committee on Long Range Soil and Water Conservation (CLSWC). Nitrogen and phosphorus mass balances: methods and interpretation. In: *Soil and water quality: An Agenda for Agriculture.* Washington DC: National Research Council National Academy Press, pp. 431-477.

Anonymous (2001) *The integration of Biodiversity into National Environmental Assessment Procedures.* Biodiversity support progaramme UNDP/UNEF/GEF.

Anonymous (2008) *African Development Bank group. Cameroon evaluation of bank group assistance to the agriculture and rural development sector, 1996-2004.* Operations Evaluation Department (OPEV), 26th February 2008.

Bacon, C. M., Getz, C., Kraus, S., Montenegro, M. & Holland K. (2012) The social dimensions of sustainability and change in diversified farming systems. *Ecology and Society* 17(4): 41.

Balmford, A., Bruner, A., Cooper, P., Costanza, R., Farber, S., Green, R.E., Jenkins, M., Jefferiss, P., Jessamy, V., Madden, J., Munro, K., Myers, N., Naeem, S., Paavola, J., Rayment, M., Rosendo, S., Roughgarden, J., Trumper, K. & Turner, R.K. (2002) Economic reasons for conserving wild nature. *Science* 297(5583): 950-953.

Bell, D. & Taylor, J. (2004) Conclusions: Emerging research themes. In Tayloy, J. & Bell, M. (eds) *Population Mobility and Indigenous peoples in Australasia and North America.* London: Routledge, pp. 262-267.

Bilsborrow, R. (1979) *Population Pressures and Agricultural Development in Developing Countries: A Conceptual Framework and Recent Evidence.* Paper presented at Population Association of America (PAA) Annual Meeting.

Bilsborrow, R. (1992) Population growth, internalmigration, and environmental degradation in rural areas of developing countries. *European Journal of Population* 8: 125-148.

Binns, J.A., Maconachie, R.A. & Tank, A.I. (2003) Water, land and health in urban and peri-urban food production: the case of Kano, Nigeria. *Land Degradation and Development* 14: 431-444.

Black, R., Kniveton D., Skeldon R., Coppard D., Murata A. & Schmidt-Verkerk K. (2008) *Demographics and Climate Change: Future Trends and their Policy Implications for Migration*, Working Paper T-27, Development Research Centre on Migration, Globalisation and Poverty.

Bosch, H. van den, Vlaming, J., Wijk, M.S. van, Jager, A. de, Bannink, A. & Keulen, H. van (2001) *Manual to the NUTMON methodology*. Wageningen, the Netherlands: Alterra/LEI, Wageningen University and Research centre.

Brady, N.C. (1996) Alternatives to Slash-and-Burn: A Global Imperative. *Agriculture, Ecosystems & Environment* 58: 3-11.

Braimoh, A. & Vlek, P. (2005) Land-cover change trajectories in northern Ghana. *Environmental Management* 36(3): 356-373.

Briassoulis, H. (2000) *Analysis of Land Use Change: Theoretical and Modelling Approaches*. The webbook of regional science (www.rri.wvu.edu/regscweb.htm). Regional Research Institute, West Virginia University.

Bray, J.R. (1960) The chlorophyll content of some native and managed plant communities in central Minnesota. *Can. J. Bot.* 38: 313-333.

Bray, J.R. (1961) An estimation of a minimum quantum yield of photosynthesis based on ecological data. *Pl. Physiol* 36: 371-373.

Brougham, R.W. (1960) The relationship between the critical leaf area, total chlorophyll content, and maximum growth rate of some pasture crops. *Ann. Bot. (N.S.)* 24: 463-474.

Brown, L.R. (1995) *Who will feed china? Wake-Up call for a small planet*. New York: The World Watch Environmental Alert Series.

Bryant, C.R., Russwurm, L.H. & McLellan, A.G. (1982) *The City's Countryside: Land and its Management in the Rural-Urban Fringe*. London and New York: Longman.

Costanza, R., d'Arge, R., Groot, R. de, Farber, S., Grasso, M., Hannon, B., Naeem, S., Limburg, K., Paruelo, J., O'Neill, R.V., Raskin, R., Sutton, P., & Belt, V. van den (1997) The value of the world's ecosystem services and natural capital. *Nature* 387: 253-260.

Daily, G.C. (eds) (1997) *Nature's services: societal dependence on natural ecosystems*. Washington, DC: Island Press.

Das, K.K., Ravan, S.A., Negi, S.K., Jain, A. & Roy, P.S. (1996) Forest cover monitoring using Remote Sensing and GIS – A case study in Dhaulkhand range of Rajaji National Park, Uttar Pradesh. Photonirvachak. *Jour. Ind. Soc. Remote Sensing* 24(1): 33-42.

Devendra, C., & Thomas, D. (2002) Smallholder Farming Systems in Asia. *Agricultural Systems* 71: 17-25.

De Bruijn, M. & Dijk, H. van (2003) Changing population mobility in West Africa: Fulbe pastoralists in Central and South Mali. *African Affairs* 102: 285-307.

De Groot, R.S. (1992) *Functions of Nature, Evaluation of Nature in Environmental Planning, Management and Decision Making.* Wolters-Noordhoff, Groningen, the Netherlands.

De Haas, H. (1998) Socio-Economic Transformations and Oasis Agriculture in Southern Morocco. In: Haan, L. de & Blaikie, P. (eds) *Looking at Maps in the Dark. Directions for Geographical Research in Land Management and Sustainable Development in Rural and Urban Environments of the Third World.* Netherlands Geographical Studies 240, pp. 65-78. Utrecht/Amsterdam: KNAG/FRW, UvA.

De Haas, H. (2009) *Mobility and Human Development.United Nations Development Programme.* Human Development Research paper 2009/01.

De Jong, G.F. & Gardner, R.W. (1981) Migration Decision Making: Multidisciplinary Approaches to Microlevel Studies in Developed and Developing Countries. In: Jong, G.F. de & Gardner, R.W. (eds) *Studies in Developed and Developing Countries.* New York: Pergamon Press.

Dixon, J., Gulliver, A., Gibbon, D., & Hall, M. (2001) *Farming Systems and Poverty: Improving Farmers' Livelihoods in a Changing World.* FAO.

Earl, H.J. & Davis, R.F. (2003) Effect of drought stress on leaf and whole canopy radiation use efficiency and yield of maize. *Agron. J.* 95: 688–696.

Earl, H.J. & Tollenaar, M. (1998) Relationship between thylakoidelectron transport and photosynthetic CO_2 uptake in leaves of three maize (Zea mays L.) hybrids. *Photosynth. Res.* 58: 245-257.

EEA (2006) *How much bioenergy can Europe produce without harming the environment?* EEA Report No 7/2006. European Environment Agency.

Ewel, J. (1986) Designing agricultural systems for the humid tropics. *Annual Review of Ecology and Systematics* 17: 245-271.

Fotsing, J.M. (1992) Stratégies paysannes de gestion de terrains et de LAE en pays Bamiléké Ouest Cameroun. *Bull. Réseau Erosion* 12: 241-254.

Fox, J. (2000) How Blaming 'Slash and Burn' Farmers is Deforesting Mainland Southeast Asia. *Asia Pacific Issues* 47: 1-8.

Garty, J., Tamir, O., Hasid, I. Eschel, A., Cohen, Y., Kamieli, A. & Orlovsky, L. (2001) Photosynthesis, chlorophyll integrity, and spectral reflectance in lichens exposed to air pollution. *J. Environ. Qual.* 30: 884-893.

Geertz, C. (1963) *Agricultural Involution.The Processes of Ecological Change in Indonesia.* University of California Press Berkeley.

Geist, H.J. & Lambin, E.F. (2001) *What drives deforestation? A meta-analysis of proximate and underlyingcauses of deforestation based on subnatinal case study evidence.* LUCC Report Series 4, pp. 116.

Ghosh, S., Sen, K.K., Rana, U., Rao, K.S. & Saxena, K.G. (1996) Application of GIS for Land-Use/Land- Cover change analysis in a Mountainous Terrain, Photonirvachak. *Jour. Ind. Soc. Remote Sensing* 24(3): 193-202.

Goldring, L. (2004) Family and collective remittances to Mexico: A multi-dimensional typology. *Development and Change* 35: 799-840.

Harwood, R.R. (1996) Development Pathways Toward Sustainable Systems Following Slash-and-Burn. *Agriculture, Ecosystems and Environment* 58: 75-86.

Henry, S., Schoumaker, B. & Beauchemin, C. (2004) The Impact of Rainfall on the First Out-Migration: A Multi-level Event-History Analysis in Burkina Fas., *Population and Environment* 25(5): 423-460.

Hoffmann, M., Hilton-Taylor, C., Angulo, A., Böhm, M., Brooks, T.M., Butchart, S.H., Carpenter, K.E., Chanson, J., Collen, B., Cox, N.A., Darwall, W.R., Dulvy, N.K., Harrison, L.R., Katariya, V., Pollock, C.M., Quader, S., Richman, N.I., Rodrigues, A.S., Tognelli, M.F., Vié, J.C., Aguiar, J.M., Allen, D.J., Allen, G.R., Amori, G., Ananjeva, N.B., Andreone, F., Andrew, P., Aquino, Ortiz, A.L., Baillie, J.E., Baldi, R., Bell, B.D., Biju, S.D., Bird, J.P., Black-Decima, P., Blanc, J.J., Bolaños, F., Bolivar, G.W., Burfield, I.J., Burton, J.A., Capper, D.R., Castro, F., Catullo, G., Cavanagh, R.D., Channing, A., Chao, N.L., Chenery, A.M., Chiozza, F., Clausnitzer, V., Collar, N.J., Collett, L.C., Collette, B.B., Cortez, Fernandez, C.F., Craig, M.T., Crosby, M.J., Cumberlidge, N., Cuttelod, A., Derocher, A.E., Diesmos, A.C., Donaldson, J.S., Duckworth, J.W., Dutson, G., Dutta, S.K., Emslie, R.H., Farjon, A., Fowler, S., Freyhof, J., Garshelis, D.L., Gerlach, J., Gower, D.J., Grant, T.D., Hammerson, G.A., Harris, R.B., Heaney, L.R., Hedges, S.B., Hero, J.M., Hughes, B., Hussain, S.A., Icochea, M.J., Inger, R.F., Ishii, N., Iskandar, D.T., Jenkins, R.K., Kaneko, Y., Kottelat, M., Kuzmin, S.L., La Marca, E., Lamoreux, J.F., Lau, M.W., Lavilla, E.O., Leus, K., Lewison, R.L., Lichtenstein, G., Livingstone, S.R., Lukoschek, V., Mallon, D.P., McGowan, P.J., McIvor, A., Moehlman, P.D., Molur, S., Muñoz, Alonso, A., Musick, J.A., Nowell, K., Nussbaum, R.A., Olech, W., Orlov, N.L., Papenfuss, T.J., Parra-Olea, G., Perrin, W.F., Polidoro, B.A., Pourkazemi, M., Racey, P.A., Ragle, J.S., Ram, M., Rathbun, G., Reynolds, R.P., Rhodin, A.G., Richards, S.J., Rodríguez, L.O., Ron, S.R., Rondinini, C., Rylands, A.B., Sadovy de Mitcheson, Y., Sanciangco, J.C., Sanders, K.L., Santos-Barrera, G., Schipper, J., Self-Sullivan, C., Shi, Y., Shoemaker, A., Short, F.T., Sillero-Zubiri, C., Silvano, D.L., Smith, K.G., Smith, A.T., Snoeks, J., Stattersfield, A.J., Symes, A.J., Taber, A.B., Talukdar, B.K., Temple, H.J., Timmins, R., Tobias, J.A., Tsytsulina, K., Tweddle, D., Ubeda, C., Valenti, S.V., Dijk, P.P. van, Veiga, L.M., Veloso, A., Wege, D.C., Wilkinson, M., Williamson, E.A., Xie, F., Young, B.E., Akçakaya, H.R., Bennun, L., Blackburn, T.M., Boitani, L., Dublin, H.T., da Fonseca, G.A., Gascon, C., Lacher, T.E. Jr, Mace, G.M., Mainka, S.A., McNeely, J.A., Mittermeier, R.A., Reid, G.M., Rodriguez, J.P., Rosenberg, A.A., Samways, M.J., Smart, J., Stein, B.A. & Stuart, S.N. (2010). The Impact of Conservation on the Status of the World's Vertebrates. *Science* 330(6010): 1503-1509.

Jiang, Y., & Huang, B. (2000) Effects of drought or heat stress alone and in combination on Kentucky bluegrass. *Crop Sci.* 40: 1358-1362.

Joshi, J., Schmid, B., Caldeira, M.C., Dimitrakopoulos, P.G., Good, J., Harris, R., Hector, A., Huss-Danell, K., Jumpponen, A., Minns, A., Mulder, C.P.H., Pereira, J.S., Prinz, A., Scherer-Lorenzen, M., Siamantziouras, A.-S.D., Terry, A.C., Troumbis, A.Y. & Lawton, J.H. (2001) Local adaptation enhances performance of common plant species. *Ecology Letters* 4(6): 536-544.

Kay, M., Stephens, W. & Carr, M.K.V. (1985) Prospects for small-scale irrigation in sub-Saharan Africa. *Outlook on agriculture* 14(3): 115-121.

Kelly, R.L. (1992) Mobility/Sedentism.Concepts: Archaelogical Measures and Effects. *Annual Review of Anthropology* 21: 43-66. In: Taylor, J. & Bell, M. (eds) *Population Mobility and Indigenous peoples in Australasia and North America.* London: Routledge, pp. 117-135.

Keys, E. & McConnell, W.J. (2005) Global change and the intensification of agriculture in the tropics. *Global Environmental Change-Human and Policy Dimensions* 15: 320-337.

Khan, M.L., Menon S. & Bawa K.S. (1997) Effectiveness of the protected area network in biodiversity conservation: a case study of Meghalaya state. *Biodiv. Conserv.* 6: 853-868.

Kleinman, P.J.A., Pimentel, D. & Bryant, R.B. (1995) The Ecological Sustainability of Slash-and-Burn Agriculture. Agriculture, *Ecosystems & Environment* 52: 235-249.

Kleinman, P.J.A., Bryant, R.B. & Pimentel, D. (1996) Assessing Ecological Sustainability of Slash-and-Burn Agriculture through Soil Fertility Indicators. *Agronomy Journal* 88: 122-127.

Lacroix, T. (2005) *Les réseaux marocains du développement: Géographie dutransnational et politiques du territorial.* Paris: Presses de Sciences Po.

Lambin, E., Geist, H. & Lepers, E. (2003) Dynamics of land-use and land-cover change in tropical regions. *Annual Review of Environment and Resources* 28: 205-241.

Levin, S.A. (ed.) (2000) *Encyclopedia of Biodiversity*, Five-Volume Set, 2nd edition. Academic Press. 4666 pp.

Lewis, J.R. (1986) International Labour Migration and Uneven Regional Development in Labour Exporting Countries. *Tijdschrift voor Economische en Sociale Geografie* 77: 27-41.

Lipton, M. (1980) Migration from the rural areas of poor countries: The impact on rural productivity and income distribution. *World Development* 8: 1-24.

Loreau, M. (2000) Biodiversity and ecosystem functioning: recent theoretical advances. *Oikos* 91: 3-17.

May, R.M. (1973) *Stability and Complexity in Model Ecosystems.* Princeton University Press.

Millennium Ecosystem Assessment (MA) (2005) *Ecosystems and Human Well-Being: Current State and Trends.* Washington, DC: Island Press.

Matson, P.A., Parton, W.J., Power, A.G. & Swift, M.J. (1997) Agricultural intensification and ecosystem properties. *Science* 277: 504-509.

Meisinger, J.J. & Randall, G.W. (1991) Estimating Nitrogen budgets for soil-crop systems. In: Follett, R.F., Keeney, D.R. & Cruse, R.M. (eds) *Managing nitrogen for groundwater quality and farm profitability.* Wisconsin: SSSA Madison, pp. 85-124.

Meyer, W.B. & Turner II, B.L. (1992) Human Population, Growthand Global Landuse/cover Change. *Annu. Rev. Ecol. Syst.* 23: 39-61.

Mukherji, S. (2001) *International Mobility Field Theory: Extension of The Mobility Field Theory for Linking Internal and International Migration.* 24th IUSSP Conference, Salvador, Brazil.

Munters, P.J.A.L. (1997) *The Dutch Manure Policy: MINAS (Nutrient accounting system).* Report from Dutch Dept. of Agriculture of the Ministry of Agriculture, Nature Management and Fisheries.

Newbold, B. (2004) Data sources and Issues for the Analysis of Indigenous People's Mobility. In: Oglethorpe J., Ericson J., Bilsborrow R.E. & Edmond J. (2007) *People on the Move: Reducing the Impacts of Human Migration on Biodiversity.* Washington, D.C.: World Wildlife Fund and Conservation International Foundation.

Nielsen, U., Mertz, O. & Noweg, G.T. (2006) The Rationality of Shifting Cultivation Systems: Labor Productivity Revisited. *Human Ecology* 34: 210-218.

Parnwell, M. (1993) *Population Movements in the Third World.* London: Routledge.

Piguet, E. (2008) *Climate Change and Forced Migration,* Research Paper no. 153. Geneva: UNHCR.

Pike, J.G. (1968) *Malawi: A Political and Economic History.* London, Pall Mall Press.

Pike, J.G. & Rimmington, G.T. (1965) *Malawi. A geographical study.* London: O.U.P., 229 p.

Powel, A.G. (2010) Ecosystem services and agriculture: tradeoffs and synergies. *Phil. Trans. R. Soc. B.* 365: 2959-2971.

Priess, J.A., Koning, G.H.J. de & Veldkamp A. (2001) Assessment of interactions between land use change and carbon and nutrient fluxes in Ecuador. *Agric Ecosyst Environ* 85: 269-279.

Quinn, Michael A. (2006) Relative Deprivation, Wage Differentials and Mexican Migration. *Review of Development Economics* 10: 135-153.

Ramankutty, N. (2010) Agriculture and forests – recent trends, future prospects. In: *Linkages of sustainability.* MIT Press, pp. 11-31.

Ratha, D. & Shaw, W. (2007) *South-South Migration and Remittances,* Washington D.C.: The World Bank Working Paper no. 102.

Reid, W.V. (1992) Tropical Deforestation and Species Extinction. In: Whitmore, T.C. & Sayer, J.A. (eds) How Many Species Will There Be? London: Chapman & Hall, pp. 55-73.

Rubenstein, H. (1992) Migration. Development and Remittances in Rural Mexico. *International Migration* 30.

Ryan, P.A. (1999) *The use of revegetated areas by vertebrate fauna in Australia: a review.* Shepparton, Victoria, Australia: GBCMA.

SID (Society for International Development) (2002) *Declaration of The Hague on the Future of Refugee and Migration Policy.* The Hague: SID Netherlands Chapter.

Simmel, G. (1990) *The Philosophy of Money,* 2nd ed., trans. Bottomore, T. & Frisby, D. London: Routledge.

Sørensen, T.A. (1948) A method of establishing groups of equal amplitude in plant sociology based on similarity of species content, and its application to analyses of vegetation of Danish commons. *Biol. Skr., Kongl. Danske Videnske. Selsk* 5: 1-34.

Stark, O. & Taylor, J.E. (1989) Relative Deprivation and International Migration. *Demography* 26(1): 1-14.

Sunderlin, W. & Pokam, J. (1998) *Economic crisis and forest cover change in Cameroon: the roles of migration, crop diversification, and gender division of labor.* Unpublished manuscript. Bogor: Center for International Forestry Research, .

Swindell, K. (1984) Farmers, traders and labourers: dry season migration from North-West Nigeria. *Africa* 54: 3-18.

Swindell, K. & Iliya, M.A. (1999) Making a profit, making a living: commercial food farming and urban hinterlands in North-West Nigeria. *Africa* 69: 386.

Swinton, S.M., Lupi, F., Robertson, G.P. & Hamilton, S.K. (2007) Ecosystem services and agriculture: cultivating agricultural ecosystems for diverse benefits. *Ecol. Econ.* 64: 245-252.

Tacoli, C. (2002) *Changing rural-urban interactions in sub-Saharan Africa and their impact on livelihoods.* Working Paper Series on Rural-Urban Interactions and Livelihood Strategies. London: International Institute for Environment and Development (IIED).

Taylor, E. (1984) Egyptian migration and peasant wives. *Merip reports* 124: 3-10.

Tilman, D. (1996) Biodiversity: population versus ecosystem stability. *Ecology* 77: 350-363.

Tilman, D., Lehman, C. & Thomson, K.T. (1997) Plant diversity and ecosystem productivity: theoretic al considerations. *Proc. Natl Acad. Sci. USA* 94: 1857-1861.

Tilman, D., Reich, P.B., Knops, J., Wedin, D., Mielke, T. & Lehman, C. (2001) Diversity and productivity in a longterm grassland experiment. *Science* 294: 843-845.

Tilman, D., Cassman, K.G., Matson, P.A., Naylor, M. & Polasky, S. (2002) Agricultural sustainability and intensive production practices. *Nature* 418: 671-677.

Trung, N.H., Tri, L.Q., Van Mensvoort, M.E.F. & Bregt, A. (2006) Comparing Land-Use Planning Approaches in the Coastal Mekong Delta of Vietnam. In: Hoanh, C. T., (eds) *Environment and Livelihoods in Tropical Coastal Zones: Managing Agriculture-Fishery-Aquaculture Conflicts.* UK: CABI Publishing.

Turner II, B.L., Moss, R.H. & Skole, D.L. (1993) *Relating Land Use and Global Land-Cover Land Change.* IGBP Report No. 34 and HDP Report No. 5. Stockholm: IGBP/HDP.

Turner II, B.L., Skole, D., Sanderson, S., Fischer, G., Fresco, L. & Leemans, R. (1995) *Land-Use and Land-Cover Change; Science/Research Plan.* IGBP Report no. 35; HDP Report No. 7.

United Nations University – Institute for Environment and Human Security (2005) *As Ranks of 'Environmental Refugees' Swell Worldwide, Calls Grow for Better Definition,* Recognition, support, UN Day for Disaster Reduction, 12th October [www.ehs.unu.edu/file.php?id=58].

Van den Bosch, H., De Jaeger, A. & Vlaming, J. (1998) Monitoring nutrient flows and economic performance in African farming systems (NUTMON). II. Tool development. *Agriculture, Ecosystems and Environment* 71: 49-62.

Van der Hoek, W., Ul Hassan, M., Ensink, J.H.J., Feenstra, S., Raschid-Sally, L., Munir, S., Aslam, R., Ali, N., Hussain, R. & Matsuno, Y. (2002) *Urban wastewater: a valuable resource for griculture. A case study from Haroonabad, Pakistan*. Research Report 63. Colombo: International Water Management Institute (IWMI).

Waggoner, P.E. & Ausubel, J.H. (2002) A framework for sustainability science: a renovated IPAT identity. *Proceedings of the National Academy of Sciences* 99(12): 7860-7865.

Watson, C.A. & Stockdale E.A. (1997) Using nutrient budgets to evaluate the sustainability of farming systems. *Newsletter of the European Network on Organic Farming* 5: 16-19.

Ying, J., E.A., L. & Tollenaar, M. (2002) Response of leaf photosynthesis during the grain-filling period of maize to duration of cold exposure, acclimation, and incident PPFD. *Crop Sci.* 42: 1164-1172.

Young, A. (1990) Agroforestry for the management of soil organic matter. In: *Organic-matter Management and Tillage in Humid and subhumid Africa*. IBSRAM Proceedings 10. Bangkok: IBSRAM, pp. 285-303.

Wells, S. (2011) *Pandora's seed: the unforeseen cost of civilization*. New York, USA: Random Hous.

Willey, R.W. (1979) Intercropping – Its importance and research needs. Part. I: Competition and Yield advantages. *Field Crops Abstr.* 32: 1-10.

Wittman, H. (2009) Reworking the metabolic rift: La Vía Campesina, agrarian citizenship, and food sovereignty. *Journal of Peasant Studies* 36(4): 805-826.

Wood, E.C., Tappan, G.G. & Hadj, A. (2004) Understanding the drivers of agricultural land use change in south-central Senegal. *Journal of Arid Environments* 59(3): 565-582.

World Resources Institute (2005) *Millennium Ecosystem Assessment*, Washington, D.C.: IslandPress.

Zhang, W., Ricketts, T.H., Kremen, C., Carney, K. & Swinton, S.M. (2007) Ecosystem services and dis-services to agriculture. *Ecol. Econ.* 64: 253-260.

Zhu, Y.Y., Chen, H., Fan, J., Wang, Y.,Li, Y., Chen, J., Fan, J.X., Yang, S., Hu, L., Leung, H., Mew, T.W., Teng, P.S., Wang, Z. & Mundt, C.C. (2000) Genetic diversity and disease control in rice. *Nature* 406: 718-722.

2 Determinants and Impacts of Human Mobility Dynamics in the Western Highlands of Cameroon

C.M. Tankou[1*], H.H. de Iongh[2], G. Persoon[3], M. de Bruijn[4], and G.R. de Snoo[2]

Submitted to International Journal of Scientific and Technology Research

1 Faculty of Agronomy and Agricultural Sciences, University of Dschang, P.O. Box 222. Dschang, Cameroon
2 Institute of Environmental Sciences, Leiden University, P.O. Box 9518, 2300 RA Leiden, The Netherlands
3 Department of Anthropology, Leiden University, P.O. Box 9518, 2300 RA Leiden, The Netherlands
4 African Studies Centre, Leiden University, P.O. Box 9555, 2300 RB Leiden, The Netherlands
* Corresponding author. cmtankou@yahoo.com; tel: (237) 77 66 03 04; fax: (237) 33 45 15 66

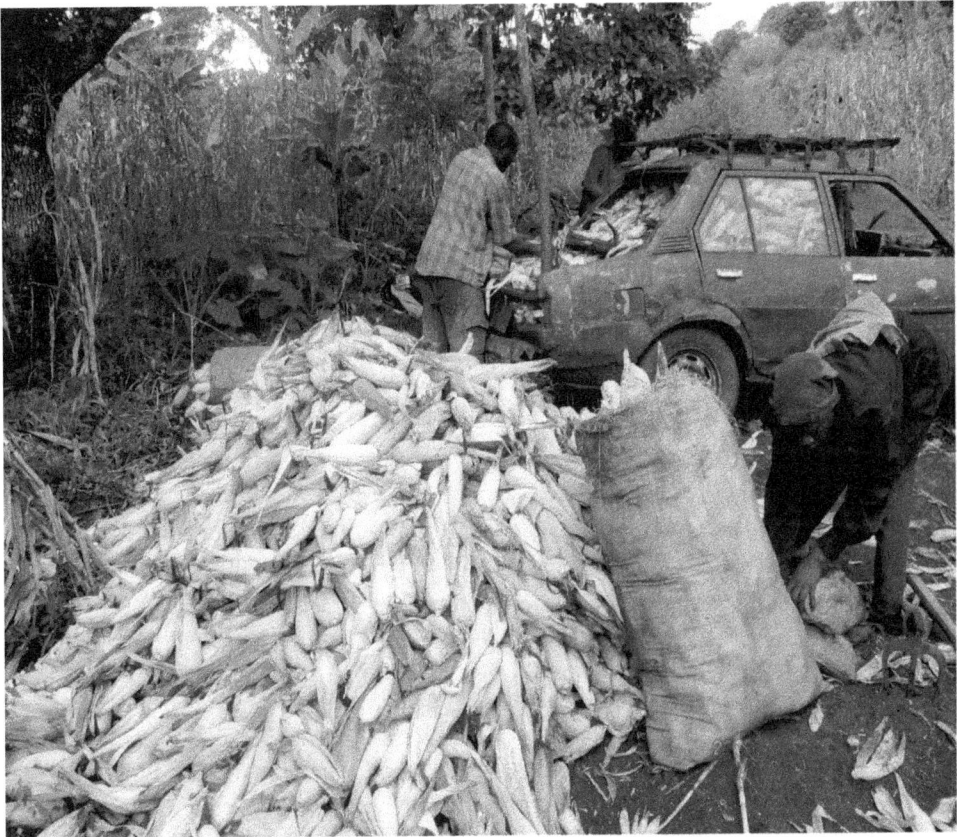

Abstract

This article analyses human mobility among the inhabitants of Cameroon's most populous region, the Western Highlands of Cameroon. In order to capture the impact of various determinants on human mobility, a comparative study was conducted through household and field surveys in three villages in the region,the study based on the systems approach. The drop in coffee prices coupled with demographic pressure was a major determinant of mobility and land-use changes in the area. Rural-to-urban migration was significantly controlled by a combination of socio-economic determinants while commuting to other rural areas for farming was triggered by the quest for microclimates suitable for the production of vegetable cash crops. Intensive land-use and a high dependence on off-farm chemical inputs had replaced the traditional long fallow system. This was found to be a threat to the sustainability of the farming system due to vulnerability to pests and erosion. On the other hand, occupational diversification triggered by urban-to-rural migration has had a far reaching effect, improving farming in rural areas and the standard of living.

Keywords

Western Highlands of Cameroon, mobility framework, determinants, impacts.

2.1   Introduction

Mobility in the rural areas has increased to a spectacular degree in most African countries because rural areas are in general no longer isolated or self-sufficient (Mabogunje, 1970; Potts, 2006). Rural-urban migration is a major contributor to urbanization in many developing countries (Tawanchai *et al.*, 2008) and an inevitable part of economic development (Kessides, 2005) driven by various determinants. The determinants so far identified that govern out-migration, include socio-politico-economic and ecological factors (Carr, 2004). Urban centres attract a significant proportion of the rural population and the wages earned there, are often remitted by migrants to their rural homelands transforming croplands and other infrastructures (Lambin *et al.*, 2001). Cameroon is an agricultural economy and the rural sector, which accounts for 30% of GDP, plays a leading role in the national economy (African Development Bank Group, 2008). Rural-rural as well as urban-rural types of mobility are radically changing the natural resources, socio-economic, demographic and development profile of the Western Highlands of Cameroon (WHC), with far reaching implications for its agricultural-based asset. However, very little empirical data exists for this region which is the major food-basket of the country. Originally, the area had a significant forest cover, but due to human mobility and farming activities, it is now dominated by different types of humid savannah vegetation (Dongmo, 1984) that reduce the production potential of the area. A lot of effort is thus required to shape the livelihood of the WHC. The socio-economic concerns on the impacts of demographic pressure have been addressed to some extent (Dongmo, 1984; Scott, 1980; Sunderlin *et al.*, 2000). Human movements in the rural areas of this region, motivated by the urge for better-quality cropland coupled with the modification of cultivation techniques in the rural environment, are of late development and require some research attention.

Though rural-rural migration involves huge numbers of people (Achanfuo-Yeboah, 1993), and the migrants are the key in population-environment relationships (Carr & Bilsborrow, 2001), the subject is widely neglected and greatly under-researched. Case studies further suggest that Africa and Asia remain at an earlier stage of migration, in which rural-rural migration dominates, whereas urban-urban migration is dominant in Latin America (Bilsborrow & Carr, 2001). From an ecological perspective, rural-rural migration is of eminent importance because of its increasing impact on the rural landscape. Less effort has also been made to research urban-rural migration, despite studies highlighting the continuing links of urban migrants with their home areas and their eventual return to villages for retirement (Peil & Sada, 1984). Information on the determinants of the various types of mobility is primary to understanding and contributing to their evolution. This paper is thus intended to contribute to this hitherto neglected field with reference to the WHC.

Mobility refers to all forms of territorial movement by people at different spatial and temporal scales. It could denote short term mobility between different dwelling sites (Kelly, 1992), or long term mobility between different areas (Kelly, 1992). Mobility and commuting decisions have been shown to emphasize the mutual dependency between migration and commuting (Zax, 1994). While the relationship between the two forms of mobility is theoretically fairly well established and appreciated, empirical work in this field has mainly concentrated on either migration or commuting (Kent *et al.*, 2003). Rural poverty in Cameroon was exacerbated in the early '90s by the devaluation of the CFA franc and slumping coffee and cocoa prices in the world market, until then the major income generators (ASB, 2003). Cash crops, which had earned 123 billion FCFA for rural households in 1984-85, only generated 6.3 billion FCFA in 1992-93 (Amin & Dubois, 1999). Commuting to farms at longer distances due to land scarcity started gaining ground when the production and export of cocoa and coffee declined precipitously after 1989, in response to the cutting of producer prices and subsidies by the government (FAOSTAT, 2010). Thus, currency devaluation exacerbated by changes in the market price of coffee, in addition to the draconian public sector wage cuts in 1993, triggered the revolution in rural farming practices, characterized by the substitution of the low-valued annual cash crop (coffee) cultivation with bi or tri-annual vegetable crop production systems in the WHC (Gubry & Lamlenn, 1996). Cool season crops are the leading cash crops in the WHC and are better adapted to tropical highlands. Commuting has thus been quasi directional in the WHC, resulting in the exploitation of higher altitudes with the appropriate microclimate for vegetable cash crop production. Migration and especially urban to rural types, has had a significant effect on the diversification of income opportunities in the rural milieu. Movements within the rural areas or between rural and urban areas, involves trade-offs for both those who move and those who stay.

The push-pull theory has been suggested as the main determinants of migration (Lee, 1966), while most researchers have recognized the overriding importance of economic motives such as the rural-urban income disparity (Eicher et al., 1970), as a significant determinant to trigger migration. Responing to land scarcity by adaptating the agricultural system to increase yield has been proposed as an important determinant of mobility (Dasgupta et al., 2000). Such adaptations usually include both intensification and increasing commercial output (Guyer, 1997). The theory of decision making on migration proposed by Byerlee (1974) stated that the policy variables affecting the decision to migrate were influenced by monetary costs and returns relating to rural and urban incomes, education, urban-rural remittances and the labour market, in addition to psychological information or non-monetary costs and returns relating to risk and life styles. Mobility is thus motivated by a multitude of determinants.

Rural farmers in the WHC and other developing countries are not only a larger group, but also far more vulnerable because of their low and volatile income, and hence deserving of more attention. It is, therefore, necessary to carry out research that better reflects the situation and experiences of the rural agricultural sector in developing countries (Zhong et al., 2007). Some factors governing mobility in the WHC have been addressed. Dongmo (1984) found that villages in the WHC adjusted to population pressure by seasonal mobility and the creation of sub-villages, while Scott (1980) suggested that the adaptation in the WHC due to demographic pressure followed the hypothesis of Boserup (1965) implying that this stimulated the adoption of improved agricultural technologies. This study attempts to provide an in-depth analysis of not onlywhy people move, but also the implications and ramifications of mobility in the WHC. Therefore the main objective of this research is to identify the determinants of the mobility and also to quantify and conceptualize the impact of human mobility dynamics on production resources in the WHC.

The specific objectives of this study are therefore:

- to identify the determinants of migration and circular movements
- to analyse the determinants of migration and circular movements
- to identify the impact of the mobility systems on rural livelihood and the environment

2.1.1 Some terminologies

a   Mobility encompasses migration and commuting (circulation) (Kent *et al.*, 2003). Population movements fall within a wide range of categories, depending on the length of time spent away from the source, the frequency and duration of the return and may involve very different kinds of people in very different circumstances (Kelly, 1992). The temporal dimension of population movement which can either be circulatory or migratory, determines the circumstances that underpin the decision to move (Parnwell, 1993).

b   Circulation encompasses a variety of movements, usually short-term and cyclical and involving no long-standing change of residence. Circulation can be subdivided into daily, periodic, seasonal, and long-term (Gould & Prothero, 1975). Daily circulation involves leaving a place of residence for up to 24 hours. Periodic circulation may vary from one night to a year, while seasonal circulation is a type of periodic circulation in which the period is defined by marked seasonality in the physical or economic environment.

c   Migration involves a permanent or semi-permanent change of residence (Newbold, 2004). Migration and circulation are therefore just different forms of the broader phenomenon of human mobility.

d   The spatial dimension of population movements is divided into two categories, internal and international. Internal movements occur within the borders of a specific country, while moving internationally means crossing the border of one country into another. Four types of mobility can be identified in the internal movements of population: urban-to-urban (intra-urban), urban-to-rural, rural-to-urban, and rural-to rural (intra-rural) (Cohen, 2006).

e   An urban areas relates to a town or city that is freestanding, densely occupied and developed with a variety of shops and services while the rural area can be identified by low population density, extensive land use, primary economic activity and employment, and community cohesion and governance (Scott *et al.*, 2007).

f   The natives of the WHC are organized into a number of independent villages, subdivided into a great number of quarters, each with its own hereditary chief (Ouden, 1987). Thus in the framework of this article, rural-to-rural mobility implies mobility in the quarters within or between villages by members of a household, either to carry out agricultural activities or work as hired labourers. Household refers to a body of people who have a shared income and asset pool and who share the same living space and eat together.

2.1.2 Conceptual framework

Our study is based on a conceptual framework of the relationships between determinants that govern natural resource use in the traditional agricultural systems and human response in terms of mobility. This study takes a comparative approach by examining household livelihoods and mobility at different altitudinal levels within three very different villages (Bafou, Baleveng and Fongo-Tongo) in the WHC. They are located in two administrative districts of the Menoua Division where Dschang is the Headquarters and main urban centre.

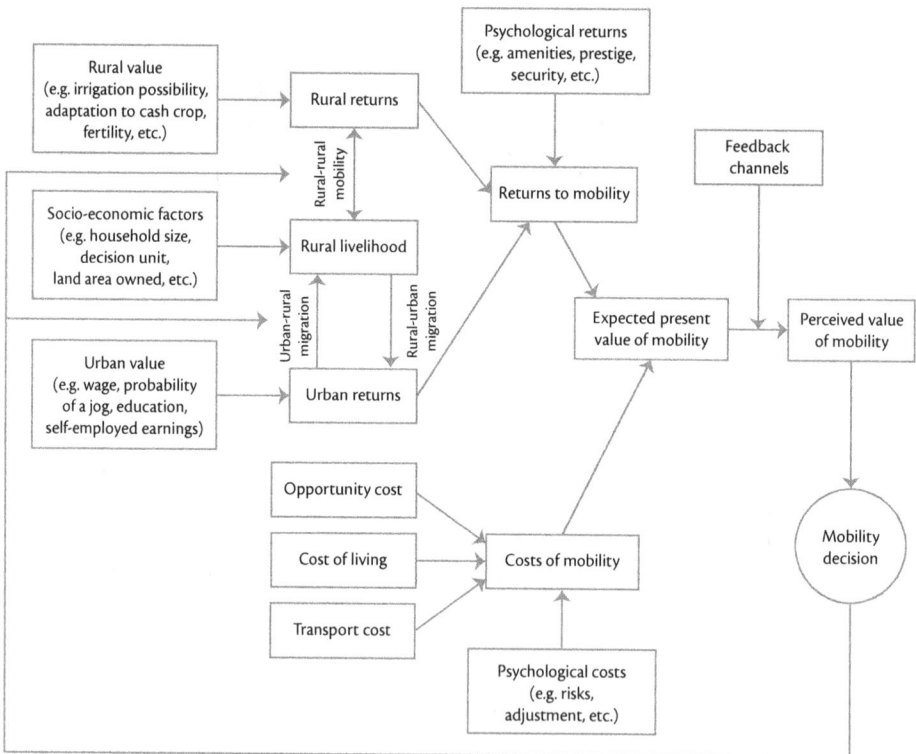

Figure 2.1
System simulation for the analysis of the migration decision and the different types of mobility in the Western Highlands of Cameroon (Adapted from Byerlee, 1974).

The mobility framework (Figure 2.1) is intended as an analytical structure for coming to grips with the complexity of mobility, understanding how it is influenced and identifying where intervention can be most effective. The assumption is, that the environment within which the system of rural-urban mobility operates is characterized by forces (economic conditions, socio-cultural conditions, transport and communications, and government policies), that have contributed to the decline in isolation and self-sufficiency of the rural areas (Mabogunje, 1970; Potts, 2006). The decision to move is governed by monetary and non-monetary or general well-being costs. The financial benefits of migration can be estimated by comparing the difference between the origin and destination incomes. Control sub-systems calibrate the flow of migrants through the system. The rural control subsystem includes the family and community while the urban control subsystem consists of opportunities for housing, employment and general assimilation into urban life. The feedback loops (positive or negative), act to favour or to disfavour the mobility decision. They can be in the form of return migration, flows of information, remittances and other forms. The adjustment mechanisms operate in rural areas to cope with the loss of migrants and in urban areas to incorporate them. The system simulation (Figure 2.1) can thus be used to conceptualize the different types of mobility (rural-to-rural, rural-to-urban and urban-to-rural) addressed in this paper, owing to its effective application in other migration studies (Kritz and Zlotnik, 1992). It identifies first the household and family members who are potential migrants when influenced by stimuli from the rural and urban environments. The components of the system are the rural environment (comprised of agro-ecosystems and natural ecosystems), the urban environment (comprised of the economy context and the socio-cultural context), the migrant, control subsystems, adjustment and feedback mechanisms. The origin in the systems thus denotes either the rural or the urban milieu where the individual resides before movement takes place, while the destination refers to the urban area in urban-rural movements or other rural areas different from the homestead in rural-rural mobility. In reality, the processes described here are not so neatly 'cut and dried': just as mobility is a dynamic process, with largely unpredictable changes in context, constraints and opportunities, as are also household strategies and activities. The drop in the price of coffee on the international market has had a major influence on mobility in the WHC.

2.2   Materials and Methods

2.2.1   Study area

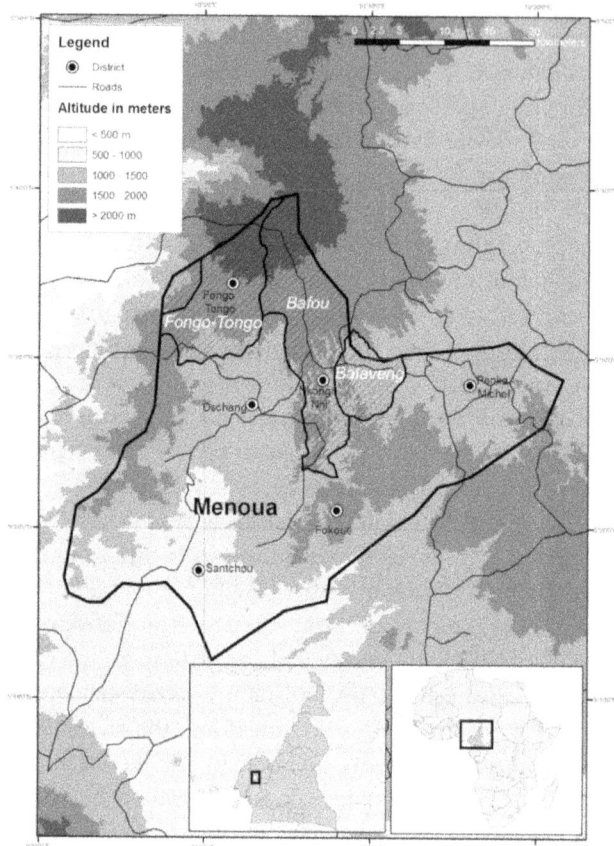

Figure 2.2
Geographical location of research site.

The study was carried out in three villages namely Bafou, Baleveng and Fon-go-Tongo all found in the Menoua Division (Figure 2-2) in Cameroon. The reason for choosing these villages was based on the fact that they had the typical characteristics of the WHC with respect to population density, altitudinal levels, distance from the urban centre, and the typical production systems common in the humid savannah. Bafou and Baleveng belong to the same district, Ndozem, while Fongo-Tongo is situated in the Fongo-Tongo district. Access to the urban centre is easier from Bafou and Baleveng, because part of Bafou and the centre of Baleveng (where the village market is located), has access to the tarmac road that links the Menoua Division to the Regional headquarters (Bafoussam), while Fongo-Tongo is more isolated, situated about 10 km away from the Dschang urban centre through difficult terrain.

Based on the classification of Kay *et al.* (1985), the WHC occupies most of the West and North-West Regions of the country with average maximum temperature of 22°C, average minimum temperature of 17°C and an annual rainfall between 1000-2000 mm falling in one long rainy season which is sufficient to grow two rain-fed crops. Fotsing (1992) described the WHC as a mountainous zone with an average altitude of about 1450 m a.s.l. characterized by granite-gneisses (metamorphosed igneous rocks) plateau in the southern lower altitudes and basaltic plateau of better agronomic quality at northern higher altitudes. Basalts weather relatively fast and chemical weathering of basalt minerals release cations such as calcium, sodium and magnesium, which give basaltic areas a strong buffer capacity against acidification. Calcium released by basalts bind with CO_2 from the atmosphere to form $CaCO_3$, thus acting as a CO_2 sink (Fotsing, 1992). The WHC was noted for a gradual degradation of the agro-sylvo-pastoral resources because of irrational exploitation. The population had however re-afforested the area to some extent with timber (*Eucalyptus sp* and *Podocarpus sp*) and fruit trees (*Cola acuminata, Dacryodes edulis, Persea americana, Mangifera indica, Canarium schweinfurthii, Spondia mombin, Citrus* spp, etc). The North-West and West Regions (Table 2.1) that make up this agro-ecological zone are the most densely populated in the country as shown by the 2005 census results (Libite, 2010). The principal cash crops in the WHC are dominated by cool-season vegetable crops after the substitution of coffee some decades ago, due to the drastic reduction in the purchase price of coffee (Sunderlin *et al.*, 2000)

Table 2-1
Population data in 2005 and land area distribution in Cameroon.

Region	Population	Area (km²)	Number of divisions	Population density (inhabitants/km²)
Adamawa	884289	63701	5	13.9
Centre	3098044	68953	10	44.9
East	771755	109002	4	7.1
Far North	3111793	34263	6	90.8
Littoral	2510363	20248	4	124
North	1687959	66090	4	25.5
North-West	1728953	17300	7	99.9
West	1720047	13892	8	123.8
South	634655	47191	4	13.4
South-West	1316079	25410	6	51.8

2.2.2 Data Collection

Data was collected from sampled households at each of three villages (Bafou, Baleveng and Fongo-Tongo) of the WHC at different altitudinal levels, from low altitudes of about 1400m to high altitudes of about 2000 m a.s.l., through surveys and structured questionnaires in 2009, 2010 and 2011. A total of 244 households participated in the study. The survey questions for the head of the households elicited both qualitative and quantitative information on the factors triggering rural-to-rural, rural-to-urban and urban-to-rural types of mobility.

2.2.3 Data Analysis

Although the mobility framework was used to synthesise the findings, the analysis itself incorporated a variety of statistical tests. The data collected from the study was subjected to both non parametric and parametric analyses based solely on the exact number of respondents for the questions concerned. The relationships between qualitative variables were analysed using the Chi square test. Treatment means that showed significant differences at the probability level of $p<0.05$ in the analysis of variance, were compared using the Student Newman-Keuls comparison test. Multiple stepwise regression analysis which generates a linear equation that predicts a dependent variable as a function of several independent variables was used to predict dependent variables responsible for movements. The relationships between the dependent variable "Number of rural-to-urban migrants" (NRUM) and the independent variables: "Fallow duration" (FD), "Size of household" (SH), "Age of head of household" (AHH), "Number of irrigable plots owned by household" (NIPH) and "Number of wetland plots owned by household" (NWPH), were evaluated through stepwise regression analyses. The last two independent variables reflected the opportunities and thus the financial comfort of the household. The equation was of the general form:

$Y = b_0 + b_1X_1 + b_2X_2 + ... + b_xX_n$ where Y was the predicted dependent variable, b_0 to b_x were partial regression coefficients, and X_1 to X_n were the independent variables (Brown, 1998). For each coefficient, the t-test determined whether the value of the coefficient was zero, and if its p-value was less than 0.05, the calculated value was considered statistically significant. Variables with p-values greater than 0.05, were sequentially excluded from the equation during stepwise regression. To assess the statistical validity of the predictive equation, we also computed the coefficient of multiple determinations (R^2 and R^2 adjusted).

2.3   Results

2.3.1   General characteristics of the study area.

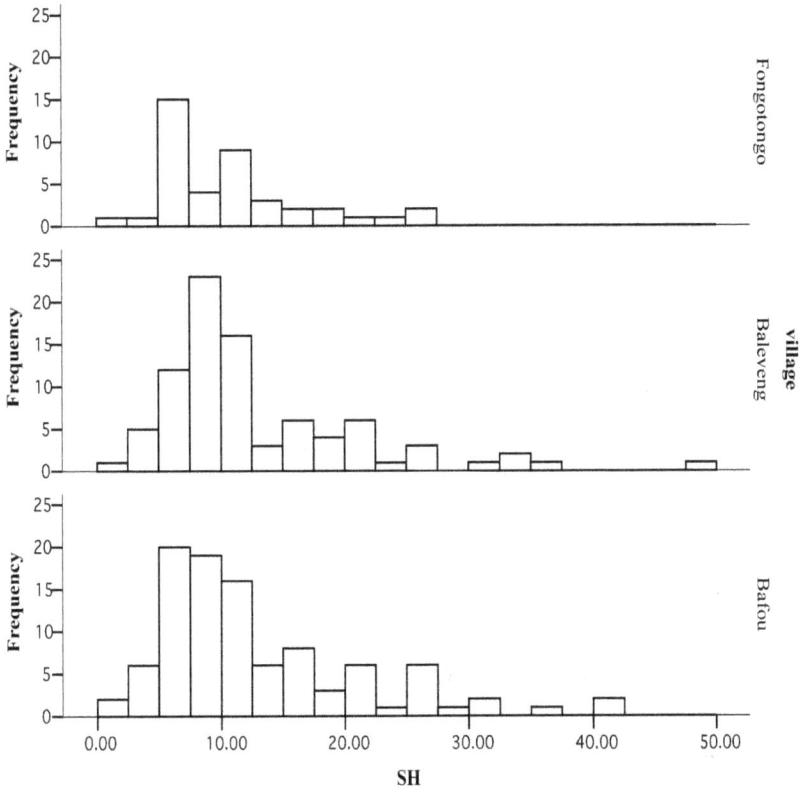

Figure 2.3
Frequency distribution of the size of the household (SH) in the study sites.

In general, the size of the household (SH) peaked at between 8 and 10 (Figure 2.3) and the age of the head of the household (AHH) peaked at 40 years, for Bafou, and 50 years for Baleveng and Fongo-Tongo (Figure 2.4).

A household consisted of a male (head of the household) and one or multiple wives and children. Female heads of households (mostly widows) were rare.

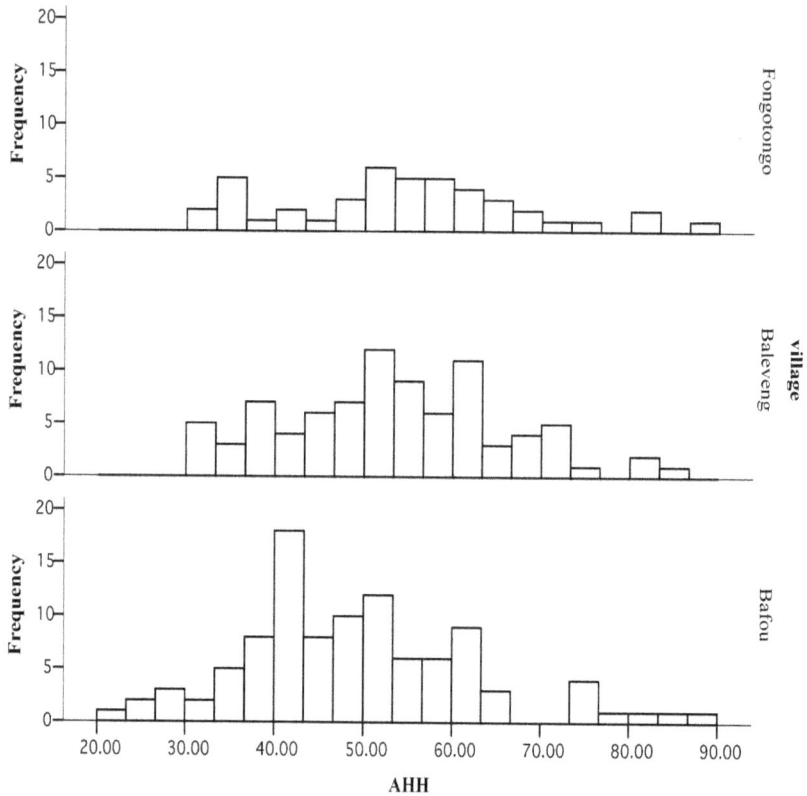

Figure 2.4
Frequency distribution of the age (years) of the household head (AHH) in the study sites.

2.3.2 Migration

In this article, two types of migration common in the WHC are discussed, viz., rural-to urban and urban-to-rural migrations. The use of mobile phones greatly alleviated the hurdles in the exchange of information in the time and space dimensions. Every respondent in this survey possessed a mobile phone and lauded its importance in both social and economic services which thus served as a feedback mechanism in the mobility system.

2.3.3 Rural-to-urban migration

Our results showed that various types of mobility with associated consequences were ongoing in this zone. Though the number of migrants to urban areas was significantly greater in Baleveng (p < 0.05) compared to Bafou, there existed no significant difference (p < 0.05) in the average numbers between Bafou and Fongo-Tongo and between Fongo-Tongo and Baleveng villages (Figure 2.5).

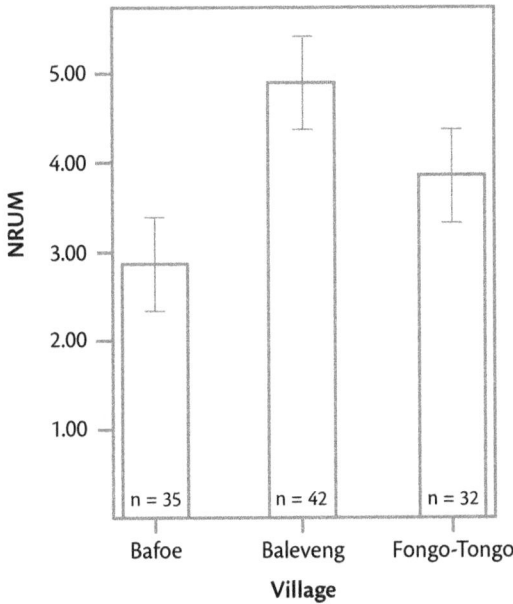

Figure 2.5
Average household
rural-to-urban mobility
(NRUM) per village.

In general, the number of people who travelled from the rural areas to the urban centres was very small compared to those who stayed behind (Figure 2.5), given that the number of members per household could be 20 or more (Figure 2.6).

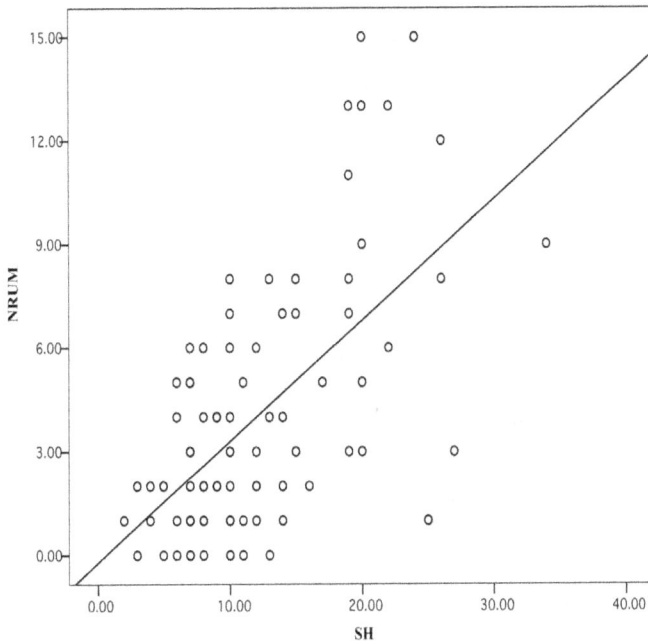

Figure 2.6
Relationship
between the size
(number of in-
habitants) of the
household (SH)
and rural-to-ur-
ban mobility
(NRUM).

There existed a strong relationship between the size of the household (SH) and the number of rural-to-urban migrants (NRUM). This was shown by the significant regression coefficient ($R^2 = 0.36^{**}$, n = 109) between the two variables (Figure 2.6). Significant influence on NRUM was also registered with the age of the head of the household (AHH) ($R^2 = 0.20^{**}$, n = 109). A stepwise analysis (Table 2.2) showed the factors that influenced out-migration.

Table 2.2

Stepwise multiple regression for rural-to-urban movement data

Model	Coefficient	Standard error	t value	P value
AHH	0.32	0.02	3.80	0.00
SH	0.53	0.05	6.18	0.00
Constant	-4.52	1.30	-3.48	0.00
R^2	0.46			
R^2 adjusted	0.45			
F-statistic	33.91			
Probability of F-statistic	0.00			

The model that combined AHH and SH was best fit for the data used to explain what influenced rural-to-urban migration. The model was defined as:

NRUM = 0.53SH + 0.32AHH - 4.52 ($R^2 = 0.46$, n = 109). Other combinations of variables that showed significant influence on NRUM were SH and the number of wet land plots owned by households (NWPH) ($R^2 = 0.38^{**}$, n = 109); and also SH and number of irrigable plots owned by households (NIPH) ($R^2 = 0.40^{**}$, n = 109),

Most of the rural-urban migrants carried out technical activities while those who moved to get married in urban areas accounted for the smallest group of migrants in the study area (Figure 2.7).

In the Batsingla sub-chiefdom of the Bafou village, a group of urban-to-rural migrants had formed an association called "*Retour au bercail*" which literally means homecoming. Most of them practiced both agriculture and their former urban occupations. In addition to their weekly activities, it was observed that they organized fortnightly meetings (on Sundays), which went on from mid-day till dusk. During these get-togethers, they shared food and drink, spent some time dancing and singing traditional songs and raised funds that were given on loan to members in need, at relatively low interest rates compared to local commercial financial institutions.

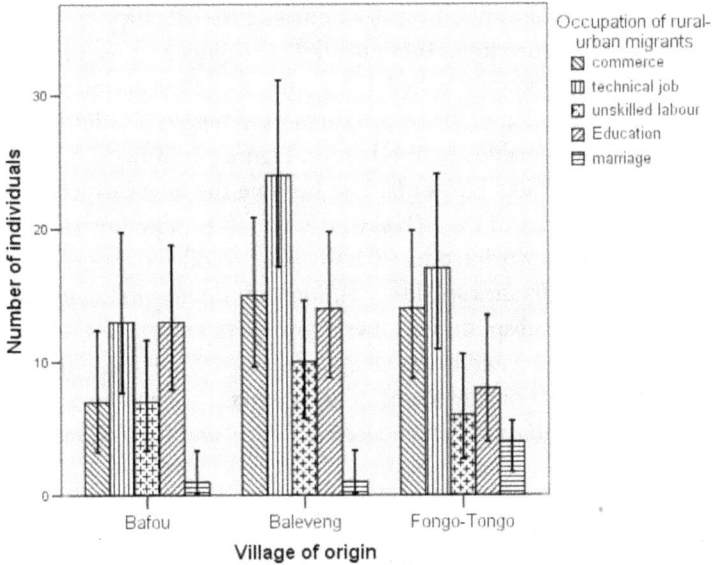

Figure 2.7
Different occupations carried out by rural-urban migrants of the study area.

2.3.4 Circular movements (Commuting)

A more crucial aspect of circular mobility had been provoked by the drop in the market price of coffee (Figure 2.8).

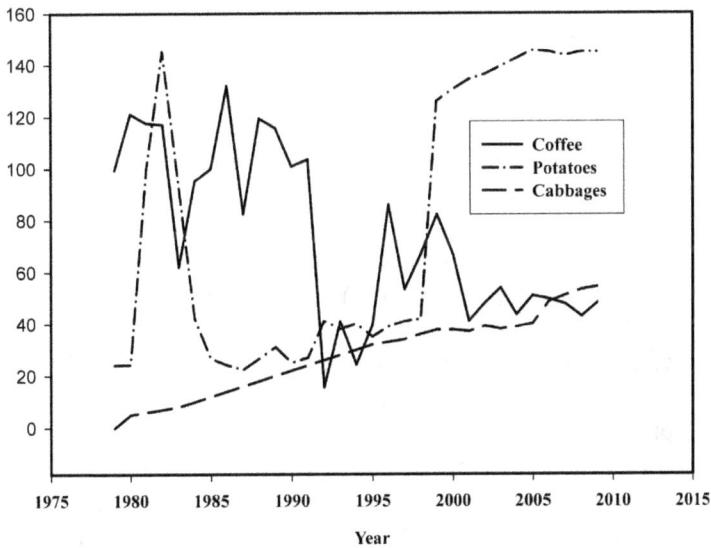

Figure 2.8
Production trend for Coffee, Potatoes and Cabbages in Cameroon (1979-2009)

The prices dropped from 0.73 euro/kg to 0.38 euro/kg for Arabica coffee and from 0.68 euro/kg to 0.24 euro/kg for Robusta coffee in 1992. The lowest production point was reached during the 1992-1994 period which marked the years of liberalization of the sector. Farmers in the WHC adjusted by substituting coffee with vegetable and food crops. Figure 2.8 shows some of the recently introduced vegetable crops that had become the new cash crops for farmers in the high altitudes of the WHC whose rate of production increased at the expense of coffee. Newly introduced cash crops included potatoes (*Solanum tuberosum*), cabbages (*Brassica* sp), leeks (*Allium ampeloprasum* var. *porrum*), carrots ((*Daucus carota* subsp. *sativus*), beetroots (*Beta vulgaris* L.), tomatoes (*Lycopersicon esculentum* L.) and beans (*Phaseolus vulgaris*), while in the low altitudes, coffee was replaced mostly by maize (*Zea mays* L.), plantains and bananas (*Musa* sp), aroids (*Xanthosoma* sp and *Colocasia* sp), and beans (*Phaseolus vulgaris*).

Table 2.3
Variation of number of farm plots at high altitude locations, number of farm plots under irrigation and maximum fallow duration with relation to villages.

Variable	Village	Means[1]
number of farm plots in high altitude areas	Bafou	2.30a
	Baleveng	0.43b
	Fongo-Tongo	0.73b
Number of farm plots under irrigation	Bafou	1.70a
	Baleveng	0.09b
	Fongo-Tongo	2.20a
Maximum fallow duration (years)	Bafou	0.83b
	Baleveng	1.81a
	Fongo-Tongo	2.27a

[1]Means followed by the same letter are not significantly different ($p < 0.05$)

Results of the analyses of variance showed significant differences ($p < 0.05$) among villages with respect to the number of farm plots acquired at the high altitude farming area per household, number of irrigable farm plots per household and the maximum fallow duration (Table 2.3), all of which constituted the main determinants of the daily circular mobility or commuting to the farms located in different quarters in the villages. The average number of farm plots occupied by households in the Bafou village (2.3) at the high altitude zone was significantly ($p < 0.05$) higher than those of Fongo-Tongo (0.73) and Baleveng (0.43). The average number of irrigable plots (based on the proximity to a water source) per household was significantly ($p < 0.05$) smaller for Baleveng (0.09) compared to Bafou (1.70) and Fongo-Tongo (2.20). An important facilitator for rural-to-rural mobility was noted to be the means of transport. Pedestrians were mostly farmers whose farms were located around the homestead while rural-rural mobile farmers travelled more by motorbike and vehicles that enabled them to reach their farms in other rural areas (Table 2.4).

Table 2.4
Relationship between the farm location with respect to the residence of the farming family and the means of locomotion of the head of the household.

Means of transportation	Farm location (%)	
	Farming within locality of homestead (n = 19)	Rural-to-rural (away from homestead) (n = 77)
Pedestrian	89.5	49.4
Use motorbike	5.3	32.5
Use vehicle	5.3	18.2
Pearson Chi Square	0.006	

The choice of markets for farm produce also accounted for commuting. Farmers who invested more in chemical and other off-farm inputs preferred selling their produce in urban markets as opposed to those who depended more on on-farm inputs (Table 2.5).

Table 2.5
Relationship between the market for farm produce and types of input used for production

Type of input used for production	Market for farm produce (%)	
	Rural market (n = 42)	Urban market (n = 20)
Chemical inputs	73.8	95
No chemical inputs	26.2	5
Pearson Chi Square		0.048

2.3.5 Mobility effects

The traditional shifting cultivation characterized by multiple cropping was noticed to be disappearing. In the Bafou village, 90% of the farmers practiced both sole and intercropping systems and 10% practiced mainly intercropping while in Baleveng, 67.3% of the farmers practiced the two systems while 32.7% practiced mainly intercropping. In Fongo-Tongo, 46.4% of the farmers practiced both systems while 71 % and 46.4% practiced mainly sole and intercropping systems respectively. Pressure on farmland and the consequent reduction in fallow period was observed in the study area (Table 2.3). In the Bafou village, fallow periods (0.83 year) were significantly ($p < 0.05$) shorter than in Baleveng (1.81 years) and Fongo-Tongo (2.27 years) villages.

There was a significant relationship between the use of chemicals (Fertilizers and pesticides) and the villages of the farmers (Table 2.6). More farmers in the Bafou village depended on off-farm inputs which included fertilizers, insecticides, fungicides and herbicides while the reverse was true for the Baleveng village.

Table 2.6
Relationship between the villages and the use of off-farm chemical inputs

Village	Use of off-farm chemical inputs	
	Yes (n = 112)	No (n = 32)
Bafou	37.5	12.5
Baleveng	31.3	53.1
Fongo-Tongo	31.3	34.4
Pearson Chi Square	0.016	

Many plant species and animal breeds were found to be either extinct or near the point of extinction in the WHC. Most farmers blamed this on the introduction of new crop production techniques. The crops that were reported to be extinct or near extinction included cultivars of cabbage (*Brassica* sp), yams (*Dioscorea* sp) and a host of spices and leafy vegetables.

Our survey results show that 85.5% of the labour force in farms far from homesteads was provided by rural-to-rural mobile waged labourers. The following sources of income were distinguished: revenue from cash crops (*Solanum* potato, cabbage, carrot, leek, tomato and beetroot), wages (regular salary or pension, hired labour), food crops, temporary jobs (e.g. carpentry, building, chain sawing, mechanics, motorbike taxi drivers etc.), petty trade, livestock, wood and remittances. The average yearly income per household in FCFA (1 € is approximately 650 FCA) ranged between: 110,000 and 420,000 in the Bafou village, 110,000 and 230,000 in the Baleveng village and 12,000 and 400,000 in the Fongo-Tongo village.

2.3.6 Urban-to-rural migration

Our survey results showed that urban-to-rural migrants were people of different walks of life which included technicians in different domains, traders and unskilled persons engaged in various activities. The urban-to-rural migrants were engaged in agricultural activities and/or their previous urban occupations for income generation (Table 2.7).

Table 2.7
Main occupations or urban-to-rural migrants of different previous urban occupations.

Main occupation after Urban-to-rural migration	Previous occupation in urban area (%)		
	Technical activity (n = 43)	Unskilled labourer (n = 10)	Commercial activity (n = 9)
Agriculture	55.8	70	77.8
Agriculture and technical activity	20.9	10	0
Agriculture and commercial activity	23.3	20	22.2

2.4 Discussion

Rural-urban migration involved the movement of rural people to urban centres particularly Dschang, which is the closest city and urban-rural migration concerned people moving from urban centres to the villages. Both types were found to be geared towards adjustments to buffer the existing socio-economic and biophysical pressures. Due to the high level of integration of rural activities into the national economy and the degree of awareness of opportunities in the rural areas through feedback from migrants, movement out of the rural areas was gaining ground. Population pressure in origin areas has been cited as a factor behind out-migration in the tropics in general (Bilsborrow & Carr, 2001) which agrees with the picture in our study sites (Figure 2.6). The positive relationship that related the size of the household with out-migration was mostly due to the burden of supporting the household population in terms of feeding, clothing, education and medical care. Rural assets in the research area such as irrigable plots and wetlands were income sources used to alleviate the burden and as such negatively encourage out-migration. It has been observed that when people have land, out-migration is attenuated (Bravo-Ureta *et al.*, 1996). A combination of social determinants was found to be responsible for the volume of rural-to-urban migration. In this study areait was both the age of the head of the household and the size of the household which represented the most influential rural control sub-system and accounted for rural-to-urban migration in search of opportunities for commerce, technical jobs, education, marriage, and the unskilled labour market in the urban areas and which were the dominant components of the environment in this system. The positive relationship between the age of the head of the household and out-migration could be related to the fact that older people would encourage household members to explore alternative income sources to improve their quality of life, given that these older people were no longer strong enough, to carry out their normal functions to support the family. These results are similar to the findings of Boyle (2004) who showed that the characteristics of a rural household that encouraged migration included a desire to improve the quality of life (with respect to education, health care, public works, entertainment, etc.). Educational achievement is mentioned as an important determinant of migration in the broader migration literature (DaVanzo, 1981; Oberai & Bilsborrow, 1984; Root & De Jong, 1991). Bongaarts (1983) proposed that marriage, fertility, adoption, mortality, migration, and divorce were proximate demographic determinants of the size of nuclear households. Macro-level institutional and political factors may determine the overall magnitude of migration, but micro-level factors play a significant role because decisions to migrate are made at the micro level and are usually household decisions (De Jong & Gardner, 1981). In this light, Bilsborrow *et al.* (1984), Findley and Li (1999) argue for an approach that integrates economic and other structural factors inherent in the context within which migration decisions are made (Bible & Brown, 1981).

Circular mobility was found to be an important type of movement in the WHC. Seasonal circular mobility was common, where family members and friends mostly from the urban areas travelled to agricultural sectors to offer their service as labourers, at the end of which they returned home. Farmers in the WHC who relied on costly off-farm inputs preferred selling their produce in urban markets where higher profits could be achieved easily and faster. These conventional farmers relied totally on chemical inputs, viz., improved planting materials, inorganic fertilizers, pesticides and herbicides for the production of the common vegetable cash crops of the area. As a result, their cost of production was relatively high compared to those who relied nearly totally on organic inputs from their farms and previous harvests as a source of planting materials. Given the fact that rural markets were not as accessible to higher bidders as urban markets, urban markets with higher bidders proved the only solution for conventional farmers wishing to make ends meet. This is in accordance with the standard microeconomics approach which hypothesizes, that people compare their earnings in their place of origin with their expected earnings at possible destinations, when making thedecision to migrate (Todaro, 1969). According to this human capital model, potential migrants will make decisions based on the economic costs and benefits of migration (DaVanzo, 1981).

Movement from the rural areas to the urban centres varied in different villages and could be influenced by the mobility channels of the system and the assets in the villages. In our study area, the highest average rural-to-urban mobility per household was recorded from the Baleveng village and the least from Bafou village. The mobility channel was greatly improved for the Baleveng inhabitants as the tarmac road leading to the Regional Headquarters cuts across the centre of the village, while the Bafou inhabitants occupied most of the favourable farmlands which explained their reluctance to move to urban centres. Migrant networks are potentially expanded by structural factors such as telecommunications and road networks and individual characteristics such as being extrovert, multilingualism, and level of education. The extent and quality of information received from friends or relatives are thus important migration destination determinants and this is confirmed by other studies (Stark & Taylor, 1991).

The results of this study suggest that young household heads with consequently young household members were less mobile because their dependents left them with little or no freedom of movement, compared to older household heads with more responsible household members capable of independent living, hence able to move about. However, the United Nations (1999) observed that the prospect of the younger generation living with their parents was becoming increasingly difficult if not impossible, as the search for employment opportunities increasingly took them to locations away from their homes and to distant lands.

Rural-rural mobility or commuting to farms in different rural neighbourhoods was highly influenced by the quest for more favourable agricultural land. The reduction in coffee prices in the early 90s had undermined the economic sustainability of the farmers in the country in general and those of the WHC in particular. The perennial nature of the coffee crop meant that adjustment to the scale of production through diversification and exiting the coffee production industry had to be slow. The absence of a national effort to assist producers accounted for the adjustments that producers were forced to make to reduce their costs, including the application of inputs and a reduced workforce in coffee production, which resulted in unemployment and migration. The reduction in labour resulted in less care being taken of trees and careless harvesting which also had adverse effects on quality which in turn led to additional pressure on average price levels.

Coffee, the original cash crop had not been location-specific but the cool-season vegetable cash crops yielded better under the low temperature conditions found at the higher altitudes in the study area and many crops per year were possible where water was unlimited. These findings coincide with Porter (1995) who noted that as temperature was such an important regulator of net photosynthetic potential, many of the most fertile areas of the tropics were at relatively high altitudes. This explained why the limited high altitude locations, irrigable and wetlands represented the destination of rural-rural mobility or commuting. AVRDC (2006) revealed that vegetables were the best resource for overcoming micronutrient deficiencies and provided smallholder farmers with a much higher income and more jobs per hectare than staple crops. The predominance of the agricultural sector and the opening up of new agricultural land were also found to be the major reasons for rural to rural migration in South Asian countries as shown by Perera (1992) where population mobility was still dominated by rural to rural migration. The results obtained in this study supported the fact that agricultural development stimulated circular movement. This was reflected in the movement of hired labourers in the rural areas. Similarly, Chapman and Prothero (1977) showed that circulation, rather than being transitional or ephemeral, was a time-honoured and enduring mode of behaviour, deeply rooted in a great variety of cultures and found at all stages of socioeconomic change. Farmers in this area being already quite satisfied with the results of vegetable crop production, subsequent increases in coffee prices could have very little effect on them, in addition to the fact that food crops had replaced most of the coffee crop. New coffee farmers found in this zone are exploiting newly bought farmlands for the cultivation of the crop.

Through the use of mobile phones, farmers in rural areas could receive or transmit secure and vital agro-produce information and produce supply instructions in real time which made their work more efficient. Most agricultural plots were small, often far from where the farmer lived. Over these distances loads of inputs and outputs had to be transported. The use of motorbikes by most farmers had helped to solve the problem of transportation in the study area. The attitude of give and take governed urban-rural remittances and rural-urban food supply. It has been shown in the study area,that in return for remittences,urban dwellers gained nearly as much, when converted into monetary terms, from the provision of food by their rural relatives.

The relatively short fallow period observed in the study area reflected the drastic change from the former shifting cultivation to the intensive land-use system shown by our results. This also explained the heavy reliance on external inputs such as fertilizers and pesticides and the practice of sole-cropping systems which had a serious impact on biodiversity. Intensive land-use was exacerbated by the exploitation of irrigable and wetlands that permitted year-round cultivation, since water is one of the most limiting factors for crop production in the area. Human pressure on land resources exacerbated by unsustainable land use practices was found to contribute towards the reduction of farm plots (Faye & Ning, 1977), fallow duration (Scott, 1980), and biodiversity (World Resources Institute, 2005), abandoned capital investment features such as terraces and irrigation (Stone, 1998; Ramakrishnan, 1992) and low productivity (Trollope & Trollope, 2004; English, 1998; Turner & Ali, 1996). Lageman (1977) observed that soil fertility, organic carbon and nitrogen declined as population pressure increased. The Global Assessment of Human-induced Soil Degradation showed that soil degradation in one form or another occurred in about 2,000 million hectares of land in the world. Water and wind erosion accounted for 84% of this damage, most of which was the result of inappropriate land management in various agricultural systems (Oldeman *et al.*, 1991).

The different types of mobility had varying effects on the livelihood of the rural population. The production of cool-season vegetables provided a better source of income compared to coffee, the former cash crop. Hired labour in the rural areas for crop production also accounted for rural-rural circular mobility in the study area. Such circular movement provided higher income at less risk than either farm production or migration (Fan & Stretton, 1980). The labour force in the rural areas was provided by inhabitants who could not move to urban areas because of the selective migration factor. Mobility has been shown to be selective (Lee, 1966) where people responded differently to the sets of plus and minus factors at the origin and destination. Selective mobility in these villages was conditional on the potential to be independent at the destination, which could be assured either by some financial backing or a technical package that could enable the creation of a business as noted by Hjort & Malmberg (2006).

The outcome of rural-rural as well as urban-rural types of mobility has had far reaching implications for the agricultural-based assets of WHC. The effects of mobility in the WHC can be grouped into: the introduction of technical expertise, loss of natural resources, diversification and the job market.
   In the study area, on-farm dependent systems could be found mainly in home-gardens many of which were found in the low altitude areas, while the higher altitudes were characterized by high off-farm input systems. Both sole and intercropping systems were practiced. The urban-control sub-system (opportunities for housing, employment and general assimilation into urban life) as proposed by Mabogunje (1970), accounted for the urban-rural migration which represented a feed-back mechanism in the mobility system. The standard of living in the rural milieu was also greatly influenced by the urban-rural migrants who introduced new professions and ideas. According to Sjenitzer & Tiemoko (2003), return migration involves the transfer of skills back to the place of origin and job improvement on the part of returning migrants. The consensus among researchers and policy makers worldwide is that poverty alleviation in the tropics can only be achieved through combining increased agricultural production with increased and diversified income for rural households (IFAD, 2001). Diversification contributes to sustain agricultural systems especially under high population densities and climatic risk (Mortimore & Adams, 1999). However, the Batsingla case (*Retour au bercail*), revealed another effect of this type of mobility. These migrants imposed a social class difference as portrayed by the segregated meetings they organized. This sug-

gested social or cultural mobility where the migrants from the urban areas claimed a higher caste owing to the experience they gained in the urban areas. The importance of both physical and cultural mobility has been proposed as aspects of a form of reflexivity that is increasingly a marker of cultural distinction and privilege in the new economy (Adkins, 2003)

Selective mobility to an extent resulted in the fact that most unskilled persons resided in the rural areas, given the job opportunities and competition in the urban centres. However the introduction of a vegetable cash crop production had opened the way for a good farm labour market, for unskilled labourers in the rural areas. In as much as diversification of enterprises increased the options for households, the introduction of modern techniques encouraged farmers to diversify, to move away from the traditional intercropping system of production to sole cropping. Intercropping has been shown to ensure efficient utilization of light and other resources, reduce soil erosion, suppress weed growth, and thereby help to maintain greater stability in crop yields. It has also been shown to guarantee greater land occupancy and consequently higher net returns (John & Mini, 2005). This explains why the sole cropping vegetable cash crop production in the WHC which was highly dependent on off-farm inputs (improved planting materials, mineral fertilizers and pesticides) was found to put land at greater risk of degradation and to be a threat to biodiversity.

2.5   Conclusion

Mobility in the WHC is guided by specific determinants and each type of movement has a specific impact on the livelihood of the population. Our results suggest that mobility does not support the unidirectional individualist and structural theories in which cause and effect relationships are much more straightforward (Bakewell, 1996). The variables determining migration to all destinations are similar, it is the relative magnitude or value of each variable and the relative vulnerability and coping options of different populations that differ. Migration networks assuage the stress associated with migration (Root & De Jong, 1991). Coffee trees take four to five years to become productive and remain productive for fifteen to twenty years thereafter. In combination with the view that coffee prices are nearly impossible to predict over a five year stretch and that farmers therefore rely on adaptible expectations when investing, this suggests potential for a coffee price cycle and a change in farmers' attitudes. However, in the WHC, an increase in coffee prices would have an insignificant effect since the majority of farmers abandoned the crop when hit the first time by low market prices. However, some new coffee farmers are emerging in the study area.

This study shows that human mobility in general can be a double-edged sword and can be a threat or an opportunity for development, as observed by Taran (2007). Mobility connects people to jobs, markets and essential services. Mobility also disconnects people from jobs and essential services as is the case with the drain of the rural workforce by the urban centres with serious consequences for agricultural production. While rural-urban migration deprives the rural areas of important human capital, return-migration brings with it knowledge, expertise and skills valuable for the socio-economic development of the rural areas. Scarcity of land in the WHC has changed the traditional extensive cultivation into an intensive sole cropping system,thus forcing the population to engage in rural-rural mobility for land expansion and to exploit vulnerable lands for farming, often without the appropriate conservation measures which cause of the loss of soil fertility and land degradation.

Acknowledgements

This research was made possible by funding from Volkswagen foundation. The authors wish to thank Maarten van 't Zelfde of the Institute for Environmental Sciences (CML), Leiden, for producing the map of the research site. We are very grateful to the extension workers of the Menoua Division who participated in the data collection and the household members who provided the useful data for the study.

References

Achanfuo-Yeboah, D. (1993) Grounding a Theory of African Migration in Recent Data on Ghana. *International Sociology* 8(2): 215-226.

African Development Bank Group (2008) *Evaluation of Bank Group Assistance to the Agricultural and Rural Development Sector. 1996-2004*. Operations Evaluation Department (OPEV).

Angermeier P.L. & Karr, J.R. (1994) Biological integrity versus biological diversity as policy directives. *Bioscience* 44: 690-697.

Amin, A.A. & Dubois, J.L. (1999) *Update of the Cameroon Poverty Profile*. Washington, DC: World Bank, 2001.

ASB (2003) *Forces driving tropical deforestation*. Nairobi, Kenya: ASB Policy Briefs November 2003. Alternatives to Slash-and-Burn.

Adkins, L. (2003) Reflexivity: Freedom or Habit of Gender. *Theory, Culture & Society* 20(6): 21-42.

AVRDC (2006) *Vegetables Matter*. AVRDC – The World Vegetable Centre. Shanhua, Taiwan.

Bell, D. & Taylor, J. (2004) Conclusions: Emerging research themes. In: Tayloy, J. & Bell, M. (eds) *Population Mobility and Indigenous peoples in Australasia and North America*. London: Routledge, pp. 262-267.

Bakewell, O. (1996) *Refugee Repatriation in Africa: Towards a Theoretical Framework?* Occasional paper 04/96. Centre For Development Studies University of Bath, UK.

Bible, D.S. & Brown, L.A. (1981) Place utility, attributes tradeoff, and choice behavior in an intra-urban migration context. *Socio-Economic Planning Sciences* 15(1): 37-44.

Bilsborrow, R.E., Oberai, A.S. & Standing, G. (eds.) (1984) *Migration Surveys in Low-Income Countries: Guidelines and Questionnaire Design.* London, Croom-Helm.

Bilsborrow, R.E. & Carr, D.L. (2001) Population, Agricultural Land Use, and the Environment in the Developing World. In: Lee, D.R. & Barrett, C.B. (eds) *Tradeoffs or Synergies? Agricultural Intensification, Economic Development and the Environment.* Wallingford, UK: CABI Publishing Co., pp. 35-56.

Bongaarts, J. (1983) *The formal demography of families and households: An overview.* IUSSP Newsletter no. 17 (January-April), pp. 27-42.

Boserup, E. (1965) *The conditions of agricultural growth: The economics of agrarian change under population pressure.* London: EartWHCan Publication.

Boyle, P. (2004) Population geography: migration and inequalities in mortality and morbidity. *Progress in Human Geography* 28(6): 767-776.

Bravo-Ureta, B., Quiroga, R. & Brea, J. (1996) Migration decisions, agrarian structure, and gender: The case of Ecuador. *Journal of Developing Areas* (4): 463-76.

Brown, C.E. (1998) *Applied multivariate statistics in Geohydrology and related sciences.* New York: Springer.

Byerlee, D. (1974) Rural-Urban migration in Africa.Theory, Policy and Research Implications. *International Migration Review* 6(4): 543-566.

Carr, D.L. (2004) A comparison of Ladino and Q'eqchi Maya land use and land clearing in the Sierra de Lacandón National Park, Petén, Guatemala. *Agriculture and Human Values* 21: 67-76.

Carr, D.L. & Bilsborrow, R.E. (2001) Population and land use/cover change: A regional comparison between Central America and South America. *Journal of Geography Education* 43: 7-16.

Chapman, M., & Prothero, R.M. (1977) *Circulation between home places and towns: A village approach to urbanization.* Paper presented at a Working Session on Urbanization in the Pacific, Association for Social Anthropology in Oceania, Monterey, California (March).

Cohen, B. (2006) Urbanization in developing countries: Current trends, future projections, and key challenges for sustainability. *Technology in Society* 28: 63-80.

Dasgupta, P., Levin, S. & Lubchenco, J. (2000) Economic pathways to ecological sustainability. *BioScience* 50: 339-345.

DaVanzo, J. (1981) Repeat migration, information costs and location-specific capital. *Population and Environment* 4: 45-73.

De Jong, G.F. & Gardner, R.W. (1981) Migration Decision Making: Multidisciplinary Approaches to Microlevel Studies in Developed and Developing Countries. In: Jong, G.F. de & Gardner, R.W. (eds) *Studies in Developed and Developing Countries.* New York: Pergamon Press.

Dongmo, J.L. (1984) Le rôle de l'homme à Travers ses activités Agricoles et Pastorales dans L'Evolution des milieux Naturels Sur les Hautes Terres de l'Ouest Cameroun. In: Kadomura, H. (ed.) *Natural and Man-Induced Environmental Changes in Tropical Africa*. Sapporo: Hokkaido University, pp. 61-74.

Eicher, C.K., Zalla, T., Kocher, J. & Winch, F. (1970) *Employment Generation in African Agriculture*. East Lansing: Institute of International Agriculture, Michigan State University.

English, J. (1998) *Malawi Impact Evaluation Report:The World Bank and the Agricultural Sector*. World Bank, Washington, DC.

FAO (2001)*The global forest resources assessment 2000 summary report*. Report No. COFO-2001/INF 5. Rome: Committee on Forestry. Food and Agriculture Organziation (FAO) of the United Nations.

FAOSTAT (2010) *Food and Agricultural Organization statistics database*.

Faye & Ning (1977) An experiment in Agrarian Restructuration and Senegalese Rural Space Planning. *African Environment* 2(4): 143-153.

Findley, A.M. & Li, F. (1999) Methodological Issues in Researching Migration. *The Professional Geographer* 51(1): 50-67.

Fotsing, J.M. (1992) Stratégies paysannes de gestion de terrains et de LAE en pays Bamiléké Ouest Cameroun. *Bull. Réseau Erosion* 12: 241-254.

Gould, W.T.S. & Prothero, R.M. (1975) Space and Time in African Population Mobility. In: Kosinski, L.A. & Mansell Prothero, R. (eds) *People on the Move: Studies on Internal Migration*. London: Methuen and Co. Ltd.

Gubry, P. & Lamlenn, S.B. (1996) Le retour du migrant au village. In: Gubry, P., Lamlenn, S.B., Ngwé, E., Tchégo, J.M., Timnou, J.P. & Véron, J. (eds) *Le retour au Village: une solution à la Crise Economique au Cameroun*. Paris, France: L'Harmattan, pp. 83-121.

Guyer, J.I. (1997) Diversity and intensity in the scholarship on African agricultural change. *Review of Anthropology* 26: 13-32.

Hjort, S. & Malmberg, G. (2006) The attraction of the rural: characteristics of rural migrants in Sweden. *Scottish Geographical Journal* 122(1): 55-75.

IFAD (2001) *Rural Poverty Report. The challenge of ending rural poverty*. Rome: IFAD, 268 p.

John, S.A. & Mini, C. (2005) Biological efficiency of intercropping in okra (Abelmoschus esculentus (L.) Moench). *Journal of Tropical Agriculture* 43(1-2): 33-36.

Kay, M., Stephens, W. & Carr, M.K.V. (1985) Prospects for small-scale irrigation in sub-Saharan Africa. *Outlook on agriculture* 14(3): 115-121.

Kelly, R.L. (1992) Mobility/Sedentism.Concepts: Archaelogical Measures and Effects. *Annual Review of Anthropology* 21: 43-66. In: Tayloy, J. & Bell, M. (eds) *Population Mobility and Indigenous peoples in Australasia and North America*. London: Routledge, pp. 117-135.

Kent, E., Lindgren, U. & Westerlund, O. (2003) Geographical Labour Mobility: Migration or Commuting? *Regional Studies* 37(8): 827-837.

Kessides, C. (2005) *The Urban Transition in Sub-Saharan Africa: Implications for Economic Growth and Poverty Reduction*. World Bank: Africa Region Working Paper Series No. 97.

Kritz, M. & Zlotnik, H. (1992) Global interactions: migration systems, processes and policies. In: Kritz, M., Lim, L. & Zlotnik, H. (eds) *International Migration Systems: A Global Approach*. Oxford: Clarendon Press.

Lambin, E.F., Turner II, B.L., Geist, H., Agbola, S., Angelsen, A., Bruce, J.W., Coomes, O., Dirzo, R., Fischer, G., Folke, C., George, P.S., Homewood, K., Imbernon, J., Leemans, R., Li, X., Moran, E.F., Mortimore, M., Ramakrishnan, P.S., Richards, J.F., Skånes, H., Steffen, W., Stone, G.D., Svedin, U., Veldkamp, T., Vogel, C. and Xu, J. (2001) The Causes of Land-Use and Cover Change: Moving beyond the Myths. *Global Environmental Change* 11: 261-269.

Lageman, J. (1977) *Traditional African Farming Systems in Eastern Nigeria: An Analysis of Reaction to Increasing Population Density*. Munich, West Germany: Weltforum Verlag.

Lee, E.S. (1966) A Theory of Migration. *Demography* 3: 47-57.

Libite, P.R. (2010) *La répartition spatiale de la population au Cameroun*. Cairo 6th ASSD.

Mabogunje, A.L. (1970) Systems Approach to a Theory of Rural-Urban Migration, *Geographical Analysis* 2(1): 1-18.

Mortimore, M. & Adams, W.M. (1999) *Working the Sahel:Environment and society in Northern Nigeria*. Routledge, London.

Newbold, B. (2004) Data Sources and Issues for the Analysis of Indigenous People's Mobility. In: Taylor, J. & Bell, M. (eds) *Population Mobility and Indigenous Peoples in Australasia and North America*. London: Routledge, pp. 117-135.

Oglethorpe J., Ericson J., Bilsborrow R.E. & Edmond J. (2007) *People on the Move: Reducing the Impacts of Human Migration on Biodiversity*. Washington, DC: World Wildlife Fund and Conservation International Foundation.

Oberai, A.S. & Bilsborrow, R.E. (1984) Theoretical Perspectives on Migration. In: Bilsborrow, R.E., Oberai A.S. & Standing, G. (eds) *Migration Surveys in Low-Income Countries*. London: Croom-Helm, pp. 14-30.

Oldeman, L.R., Hakkeling, R.T.A. & Sombroek, W.G. (1991) *World map of the status of human-induced soil degradation: An explanatory note*. Wageningen and Nairobi: International Soil Reference and Information Centre and UNEP.

Ouden, J.H.B. den (1987) In Search of Personal Mobility: Changing Interpersonal Relations in Two Bamileke Chiefdoms, Cameroon. *Journal of the International African Institute* 57(1): 3-27.

Parnwell, M. (1993) *Population Movements in the Third World*. London: Routledge.

Peil, M. & Sada P.O. (1984) *African Urban Society*. Chichester: John Wiley and Sons.

Porter, P.W. (1995) Note on Cotton and Climate: A Colonial Conundrum. In: Isaacman, A. & Roberts, R. (eds) *Cotton, Colonialism, and Social History in Sub-Saharan Africa*. London: James Currey.

Potts, D. (2006) Rural Mobility as a Response to Land Shortages: The Case of Malawi. Popul. *Space Place* 12: 291-311.

Ramakrishnan, P.S. (1992) *Shifting agriculture and sustainable development: an interdisciplinary study from North-Eastern India.* Carnforth: Parthenon Publ.

Root, B.D. & Jong, G. de (1991) Family migration in a developing country. *Population Studies* 45(2): 2212-2233.

Scott, A., Gilbert, A. & Gelan, A. (2007) *The Urban-Rural Divide: Myth or Reality?* Aberdeen AB15 8QH, UK: SERG Policy Brief Series. Macaulay Institute Craigiebuckler.

Scott, W. (1980) *Development in the Western Highlands. United Republic of Cameroon.* Yaounde: USAID/Cameroon Office of Agricultural and Rural Development.

Sjenitzer, T. & Tiemoko, R. (2003) *Do Developing Countries Benefit from Migration? A Study of the Acquisition and Usefulness of Human Capital for Ghanaian Return Migrants.* Sussex Centre for Migration Research.

Stark, O. & Taylor, J. (1991) Migration incentives, migration types: the role of relative deprivation. *Economic Journal* 101(408): 1163-1178.

Stern N. (ed.) (2006) *Stern Review on the Economics of Climate Change.* Cambridge: Cambridge University Press.

Stone, G.D. (1998) Keeping the home fires burning: the changed nature of house holding in the Kofyar homeland. *Human Ecology* 26: 239-265.

Sunderlin, W.D., Ndoye, O., Bikie, H., Laporte, N., Mertens, B. & Pokam, J. (2000) Economic crisis, small-scale agriculture, and forest cover change in southern Cameroon. *Environmental Conservation* 27(3): 284-290.

Taran P. (2007) Launch Seminar OSCE-ILO-IOM. *Handbook on Labour Migration Mediterranean.* Edition Rabat, Morocco, 12-13 December.

Tawanchai, J., Martin, P. & Trudy, H. (2008) Rural-urban migration, illicit drug use and hazardous/harmful drinking in the young Thai population. *Addiction* 103: 91-100.

Trollope, W.S.W. & Trollope, L.A. (2004) *Prescribed Burning in African Grasslands and Savannas for Wildlife Management.* Arid Lands Newsletter 55, May/June.

Todaro, M.P. (1969) A model of labour migration and urban unemployment in less developed countries. *American Economic Review* 59(1): 138-148.

Turner, B.L. & Ali, A.M.S. (1996) Induced intensification: Agricultural change in Bangladesh with implications for Malthus and Boserup. *Proceedings of the National Academy of Sciences of the United States of America* 93: 14984-14991.

United Nations (1999) *The Family and Older Persons in Bangladesh, Pakistan and Sri Lanka.* Asian Population Studies, No. 151, New York.

World Resources Institute (2005) Millennium Ecosystem Assessment, Washington, D.C., Island Press.

Zhong, F., Zhu, J. & Xie, Z. (2007) *Resource Mobility and Agricultural Trade Policy.* Proceedings at an IATRC Symposium July 8-9, 2007 Beijing, China.

Zax, J.S. (1994) When is a move a migration? *Reg. Sci. & Urban Econ.* 24: 341-60.

3

Sustainability and other determinants of smallholder farming systems in the Western Highlands of Cameroon

C.M. Tankou[1*], G.R. de Snoo[2], G. Persoon[3] and H.H. de Iongh[2]

Submitted to the Journal of Agriculture and Human values

1 Faculty of Agronomy and Agricultural Sciences, University of Dschang, P.O. Box 222 Dschang, Cameroon.

2 Institute of Environmental Sciences, Leiden University, P.O. Box 9518, 2300 RA Leiden, The Netherlands (email G.R. Snoo: snoo@cml.leidenuniv.nl; tel: +31 (0)71 527 7456; fax: +31 (0)71 527 7434; email H.H. de Iongh: iongh@cml.leidenuniv.nl; tel: +31 (0)71 527 7431; fax: +31 (0)71 527 7434)

3 Department of Anthropology, Leiden University, P.O. Box 9518, 2300 RA Leiden, The Netherlands. (Email G. Persoon: Persoonga@fsw.leidenuniv.nl; tel: +31 (0)71 527 6826; fax: +31 (0)71 527 7434)

* Corresponding author. cmtankou@yahoo.com; tel: (237) 77 66 03 04; fax: (237) 33 45 15 66

Abstract

Smallholder farming systems in the Western Highlands of Cameroon (WHC) have undergone changes in land use, productivity and sustainability. Understanding the determinants that influence the system is essential when targeting appropriate intervention strategies for improvement. A field survey was carried out in three villages in this agro-ecological zone and analysed to understand the sustainability, general characteristics of the households and other forces that drive the farming systems in this area. The impacts of farming practices on farm sustainability were used as indicators to score sustainability in our research area. The results revealed that the household characteristics were very similar across the villages while the sustainability though generally low, differed depending on the intensity of off-farm inputs in the production systems and other socio-economic factors. Sustainability had significant negative relationships with the intensity of land use, off-farm inputs, and sole cropping practice and a positive relationship with the age of the head of the household. Study of the covariance relationships among the determinants using factor analysis showed that the determinants could be grouped to indicate a number of underlying common factors influencing sustainability. The common factors were intensity of land use over space, intensity of off-farm inputs, household adjustment factors and the mobility of the household, in descending order of importance, which explained 62.15% of the total variation of sustainability in the study area. Efforts are required to improve the sustainability of the farming systems in the WHC. The adoption of well designed intercropping systems and the use of natural organic resources for plant nutrients would be of benefit and provide satisfaction to both the producers and consumers in the system.

Key words

Farming systems, sustainability, determinants, Western Highland of Cameroon.

3.1   Introduction

Many arable areas in sub-Saharan Africa, most of which are in degraded or low potential areas, have been shown to be under severe pressure to increase productivity in order to feed a rapidly growing human population (Tchabi *et al.*, 2008; Place *et al.*, 2003). The sustainability of the farming systems are negatively linked toe poor soil fertility management and continuous cropping that exacerbates soil nutrient depletion (Waithaka *et al.*, 2006), since the farming systems are usually located in heterogeneous environments too marginal for intensive agriculture and remote from markets and institutions (Tchabi *et al.*, 2008; Wolf, 1986). Most households in sub-Saharan Africa make their living from growing crops and/or keeping livestock on small plots of land which makes it a precarious and insecure way of life. Since it is not possible to increase the area under production because of demographic pressure, effective technologies are required to increase farm productivity and enhance sustainability, thereby improving the well-being of these small holders. Based on modern research, the introduction of improved technology and methods of conservation for smallholder farming, without efforts first being made to understand the determinants of the system and farmers' perceptions, are usually not effective (Isaac *et al.*, 2009; Oreszczyn *et al.*, 2010). Understanding farmers' perceptions and determinants of agriculture remain a challenge for the adoption of environmentally sustainable practice.

Agriculture is a major earner of foreign exchange for Cameroon (30% of Gross Domestic Product) and provides employment for the bulk of the population. Most of the agricultural production is by small-scale farmers of the rural areas who make up about 90% of the farming population (FAO, 2002). Farming systems in the Western Highlands of Cameroon (WHC) have evolved over time yielding both positive and negative contributions to rural welfare and livelihood. The traditional on-farm input-dependent system characterized by shifting cultivation and intercropping is no longer sustainable because factors such as socio-economic and demographic pressure have shortened the fallow period. Asa result the rural population is forced to look for other income generating farming systems, especially after the drastic drop in the market value of the original cash crop, coffee.

The evaluation of agricultural production systems is an important step in the diagnosis of the systems which will yield strategies that can be used to improve the system. These include better decision-making (lowering risks and costs), an early warning system for emerging issues, sustainability balanced with development, understanding what impacts on the systems and allowing for corrective action, identifying limits and opportunities, continuous improvement and accountability and communication (Russillo & Pintér, 2009). Very few research attempts have so far been carried out to diagnose the factors governing agricultural production in the rural areas of Cameroon and especially in the Western Highlands of Cameroon (WHC) which is considered to be the food basket of the country. Land degradation has been attributed to demographic pressure as subscribed to by the neo-Malthusian theory (Malthus, 1989) though other proponents (Tappan & McGahuey, 2007) of the Boserup's (1965) theory have argued that technological innovations follow increasing population pressure. Traditional farming systems known for preserving soil health and quality have in fact shifted to 'mining' agriculture whose duration depends only on the depletion rates of assimilable nutrients (Van der Pol, 1992). Ruthenbeng (1980) pointed out that farmlands under the traditional system were originally part of natural systems close to the "steady state" but considered unproductive in terms of human objectives. "Slash and burn" or shifting cultivation is perhaps one of the best examples of an ecological strategy to manage agriculture in the tropics (Yanni, 1996). By maintaining a mosaic of plots under cropping and some in fallow, farmers capture the essence of natural processes of soil regeneration typical of any ecological succession. By understanding the rationale of the system, a contemporary discovery, the use of "green manures", has provided an ecological pathway to the intensification of the shifting cultivation, in areas where long fallows are not possible anymore due to population growth or the conversion of forest to pasture (Flores, 1989). Hence, assessment of the factors prevailing in the systems can help to check the faults and provide a guide to alternative and more sustainable exploitative techniques.

Agriculture or farming is a high-risk business, subject not only to pests and weather but also to changes in resource availability (scarcity or deterioration), market conditions and government policies. A complex combination of stimuli, opportunities and internal adjustment mechanisms has defined different mobility routes and destinations for the rural inhabitants of the WHC, all of which have significantly impacted on the sustainability of the farming system. As a result of demographic pressure and local land tenure policy, fallows have nearly disappeared in the WHC (Floret, 1998) and land degradation has been exacerbated by the exploitation of vulnerable lands. Household characteristics and the interaction of exogenous and biophysical factors, result in highly diverse, mixed smallholder agricultural systems (Shepherd & Soule, 1998; Wopereis *et al.*, 2006). Differences among households in labour availability, resource endowments and other conditions give rise to different approaches to managing resources, even within the same region. These management differences affect the type and growth of plants, the presence and productivity of livestock, the use of fertilizers and the functioning of soil micro- and macro-fauna, which in turn influence soil fertility and the sustainability of the production system. Smallholder farming is the only option for a large proportion of the rural populations in sub-Saharan Africa. The difficulties they face is the need to strike a balance between competing needs such as maximizing labour productivity, providing themselves with a livelihood and reducing land degradation. Many farmers practice low-input subsistence farming with the aim of satisfying food requirements and basic income demands. For such systems both productivity and sustainability are at risk unless there is some use of external resources. Additionally, smallholders have to find a balance between investing in inputs for crop and livestock production, growing food for the household and generating income to buy food that cannot be grown on the farm as well as providing for health, education, and other household and social needs. The adoption of economically sustainable land management practices and technologies is constrained by a shortage of land and capital resources (Shepherd & Soule, 1998). Raising agricultural productivity in smallholder agriculture systems requires an understanding of how the complex array of farm enterprises and household socioeconomic factors relate and interact with each other.

Many factors influence the farming systems of small-scale farmers. The preoccupsions of the farmer will depend on the size of the area to be farmed, since this is the factor that determines the amount of of inputs required and isthe source of outputs. The type of crops grown is influenced by the dependency of household members, risk aversion, and the discount rate of the enterprise (Walker *et al.*, 1986). In the WHC, market oriented crops are principally sole-cropped while crops produced to feed households are generally intercropped. The age of the head of the household and the household proper influence the type of production system. As the children age and expand the family labour force, and as the household head acquires experience, production constraints are relaxed, discount rates are lowered, and risk aversion is mitigated (Walker *et al.*, 1986). The types and sources of inputs are important considerations in determining the sustainability of the systems. Systems that emulate nature and rely to a lesser extent on external inputs, just as do mature ecosystems, may provide pointers for ecologically appropriate agricultural management (Dalsgaard *et al.*, 1995). Practices that depend on non-renewable inputs and negatively contribute to natural biological processes and biodiversity show little consideration for the future generations of farmers (Rigby *et al.*, 2001). These constitute salient indicators of agricultural production.

Owing to insignificant diversification, the rural community of Cameroon in general and the WHC in particular depend nearly entirely on agricultural activities for food, feed and income. Rising demographic pressure has imposed intensive land system use over space and time and this in turn demands high amounts of off-farm inputs which seriously puts in question the sustainability of the system. According to Bergeret and Djoukeng (1993), the West region is considered to be one of the regions where agriculture is very dynamic. With 11% of the population occupying 3% of the national territory, with a density of about 200 inhabitants per square kilometre compared to the national average of 25 inhabitants per square kilometre, this region is believed to have a productive and intensive agricultural system. Most of the food and vegetable crops are exported to the big towns of the country and other neighbouring countries. However, the unregulated amounts of inputs used by farmers in this region necessitate some research on the health of the agricultural practices. This study attempts to quantify the sustainability of agricultural practices in the WHC adapted from sustainability indices developed by Rigby *et al.* (2001). The findings of this study are intended to make a contribution to the formulation of policies for sustainable agriculture development. They also provide a framework useful for the assessment of the sustainability of agricultural systems. Empirical research in the WHC addressing relationships between household structure and farm systems is limited. In addition to evaluating the sustainability of the different research sites using appropriate indicators, this study analyses the major factors influencing the farming systems in WHC by reducing a large number of inter-related variables to a few underlying factors that interact and determine the activities and performance of the agricultural system with an emphasis on sustainability. The specific objectives are:

1   to evaluate the relationship existing between agricultural production variables including sustainability score in the WHC
2   to illustrate possible interpretations of the influence of farming system determinants on sustainability
3   to identify the main constraints that influence agricultural production in the area

3.2   Materials and methods

3.2.1   Framework for assessing agricultural sustainability

The concept of sustainability lies at the heart of current debates over the use of the planet's natural resources. Agriculture is the most important user of environmental resources, including water, forests, pastures and nutrients, and its sustainability depends upon their availability (DFID, 2002). A growing interest in agricultural sustainability stems from concern about both threats to agriculture, the negative impact of agriculture on the environment, and the realization that decisions made now can have unforeseeable consequences in the future (Hansen et al., 1997). Ikerd (1993) defines sustainability as the abilityof maintain productivity and usefulness to the society in the long term, with environmentally sound, resource-conserving, economically viable, socially supportive, and commercially competitive characteristics. Sustainability is thus concerned with the need for agricultural practices to be economically viable, environmentally considerate and able to meet human food, feed and fibre needs in the long run (USDA, 1999; ATTRA, 2003) and thus integrates production and distribution (Lynam, 1994).

Sustainability takes into account economic, social and environmental concerns (Rasul & Thapa, 2003). This complex combination of interests makes it difficult to readily take a line of action to implement sustainability owing to the absence of simple diagnostic tools essential to evaluating the environmental effects of agricultural practices. The information needed for such evaluation is often difficult to obtain for financial or technical reasons (Girardin et al., 1999).

To judge the sustainability of a system it is necessary to identify a set of attributes that constitute the components of a sustainable system, to develop measurement techniques for these indicators or performance criteria, and find some way of combining them to give a broad-based, multi-factor assessment of sustainability (Spenser & Swift, 1991). Due to variations in biophysical and socioeconomic conditions, indicators used in one country are not necessarily applicable to other countries. Therefore, indicators should be location specific and constructed within the context of contemporary socioeconomic situations (Dumanski & Pieri, 1996). The relevance of the indicators to assess sustainability and their usefulness both from societal and the farmers' perspective were considered in selecting them. The indicators make use of specific farming practices backed by pertinent literature criteria commonly adopted for agricultural sustainability (Rigby *et al.*, 2001).

A lot of research has been conducted to illustrate sustainable practices. The increased use of inorganic fertilizers, insecticides and pesticides in sole cropping systems has led to the contamination of water bodies and the spread of diseases, which have adversely affected aquatic life, livestock and people (Rahman & Thapa, 1999). Enormous losses are incurred in widely planted pure stands of high-yielding varieties (IRRI, 1976) when pests develop resistance to pesticides. Pesticide misuse can kill the numerous natural enemies of pests, causing pest resurgence and infestations by formerly innocuous secondary pest species. Pesticides are also a potential hazard to humans and the environment when the developing world is ill-equipped even to monitor the extent of the problem (Bull, 1982). Moreover, many farmers and governments cannot afford the large cost of agricultural inputs. Monocrops may be less productive under tropical conditions than well-conceived polycultures (Kass, 1978). Pingali *et al.* (1991) noted that intensive rice monoculture in the rice bowls of Asia resulted in: rice paddies flooded for most of the year without adequate drying periods, increased reliance on inorganic fertilizers, asymmetry of planting schedules and greater uniformity of cultivars. Over the long run, these changes imposed significant ecological costs due to negative bio-physical impacts such as the build-up of salinity and water logging, declining soil nutrient status, increased incidence of soil toxicities, pest build-up and reduced resilience of the ecosystem to pest attack. Yield advantages of multiple cropping systems are due to the reduction of pest incidence and more efficient use of nutrients, water and solar radiation (Altieri, 2002). Increased parasitoid and predator populations, availability of alternative prey for natural enemies, decreased colonization and reproduction in pests, chemical repellents, masking feeding inhibition by odours from non-host plants, prevention of pest emigration and optimum synchrony between pests and their natural enemies are presumably important factors in efficient pest regulation in intercrops (Altieri *et al.*, 1978; Gagné, 1982; Risch, 1981). The main strategy should be that a production system should exhibit tight nutrient cycling, complex structure and enhanced biodiversity. The expectation is the establishment of agricultural mimics, like natural models, that can be productive, pest-resistant and conservative of nutrients (Ewel, 1999). Badgley *et al.* (2007) found that organic methods could produce enough food on a global per capita basis to sustain the current human population, and potentially an even larger population, without putting more farmland into production. Based on 293 examples, theyconcluded that on average, in developed countries, organic systems produce 92% of the yield produced by conventional agriculture while in developing countries organic systems produce 80% more than conventional farms. They also posited that leguminous cover crops could fix enough nitrogen to replace the amount of synthetic fertiliser currently in use. Pretty *et al.*, (2006) examined 286 projects in 57 countries and found that farmers had increased agricultural productivity by an average of 79% by adopting sustainable agricultural practices such as integrated pest and nutrient management, conservation tillage, agro-forestry, water harvesting in dry land areas, and livestock and aquaculture integration in farming systems. Pretty & Hine (2001) found that farmers had, by adopting sustainable agricultural practices, achieved substantial increases in per hectare food production - the yield increases were 50-100% for rain-fed crops, and 5-10% for irrigated crops. Other specific examples of increased yields following the application of sustainable agricultural practices have been documented (Parott & Marsden, 2002; Pretty & Hine, 2001). These include:

- In the dry lands of Burkina Faso and Niger, soil and water conservation have transformed formerly degraded lands enabling the average family to produce an annual surplus of 153 kg of cereal per year.
- In Ethiopia, sustainable agriculture has resulted in a 60% increase in crop production
- In Honduras and Guatemala crop yields have increased from 400-600 kg/ha to 2000-2500 kg/ha using green manures, cover crops, contour grass strips, in-row tillage, rock bunds and animal manures
- In Peru, Bolivia and Ecuador farmers have increased potato yields by three fold, particularly by using green manures to enrich the soil.
- In Brazil, use of green manures and cover crops increased maize yields by between 20-250%.
- In Senegal, composting systems, green manures, water harvesting systems and rock phosphate increased yields of millet and peanuts by 75-195% and 75-165% respectively

Agricultural systems that will be able to confront future challenges are those that will exhibit high levels of diversity, productivity and efficiency (Funes-Monzote, 2009) as shown in Figure 3.1.

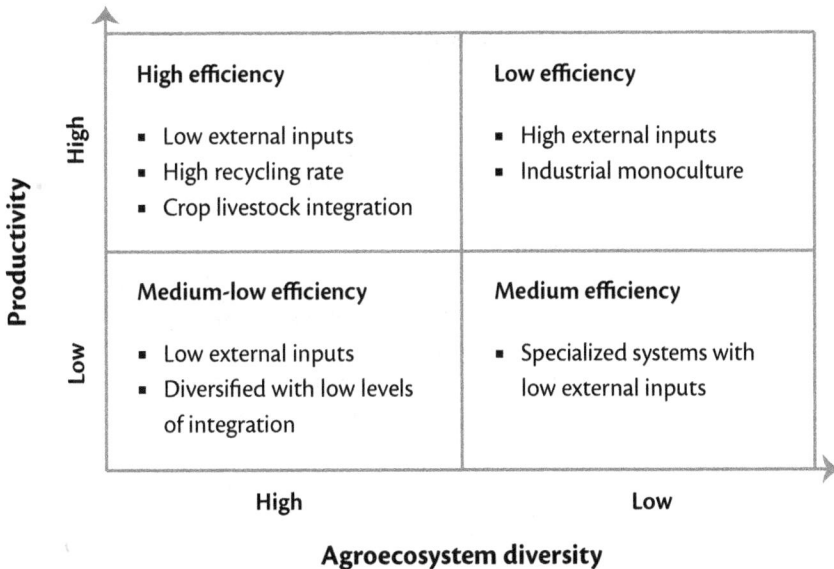

Figure 3.1
Features of green agro-ecosystems of the future: productivity, diversity, integration and efficiency (Funes-Monzote, 2009)

3.2.2 Sustainability score

Based on the principles of sustainability and the indicators outlined by Rigby *et al.* (2001), the impact of farming practices on farm sustainability was used as indicators to score the relative impact of farming practices on farm sustainability in our research area. Farm-management indicators are "raw" data that can be directly linked to activities. They can provide an early indication of likely changes which impact the environment, sometimes well before they can be measured by other indicators, such as those pertaining to soil and water quality. They can also serve as a proxy for "state" indicators where the latter are difficult or costly to monitor. Measuring farming practices is often more practical and cheaper than measuring actual changes in the environment (OECD, 2001).

Sustainability score (SUS) was estimated from the information gathered from 144 households in three villages (Bafou, Baleveng and Fongo-Tongo) found in the WHC. Information collected was based on seed source, maintenance of soil fertility, pest and disease control, crop management and weed control. The different options within each of the categories are outlined in Table 3.1.

Table 3.1

Farm practices used in the research area adapted from Rigby *et al.* (2001).

Seed source	Fertilizers	Pest/disease control	Crop manage-ment	Weed control
Impr = Improved planting material obtained off-farm	Synth = Synthetic fertilizers such as compound NPK, urea, superphosphate	Nat = Use of woodash, fallow	Rotat = rotation	Herb = chemical herbicides
Prev = Planting material obtained from previous harvest	Org = Organic fertilizers such as non-composted straw, FYM, animal dung, plant waste.	Synth = Synthetic pesticides (insecticides, fungicides, nematocides)	Inter = inter-cropping to encourage ecological diversity	C&C = Crop and Compost control (crop rotation, composting manure and crop waste to kill weed seeds)
	Comp = Composted such as organic fertilizers aerobically composted to kill pathogens			C Mgt. = Management of the crop (hand weeding or manual cultivation)

The scoring practice with respect to sustainability is given in Table 3.2 and it combines information from Table 3.1 on the different types of practices. Each farming practice was scored in absolute terms ranging from 0 to +3 points based on the criterion.

The scoring system could be interpreted positively or negatively as: 0 = no significant impact, 0.5 = marginal impact, 1 = significant impact, 2 = strong significant impact and 3 = very strong significant impact. The five categories of farm practice represent different proportions of the total number of points available. For example if a farmer depends totally on organic sources of fertilizer that accounts for +2.5 points whereas a farmer who depends totally on synthetic fertilizer would earn -8 points for that farm practice (soil fertility).

The score for each household was calculated by multiplying the total score attributed to each farm practice in Table 3.2. Hence the index values ranged between -14.5 and 23.5 depending on each household's pattern of input use in production. A linear transformation was applied to the values calculated so that the index scores ranged between 0 and 1.

Table 3.2
Scoring practices with respect to sustainability Rigby *et al.* (2001).

Farm practice	Dimension of sustainability				Total
	Minimises off-farm input	Minimises non-renewable inputs	Maximises natural biological processes	Promotes local biodiversity	
Seed sourcing					
Impr					+0
Prev	+1	+1			+2
Soil fertility					
Synth	-1	-1	-1		-3
Org	+1			+1	+2
Comp	+1	+1		+2	+4
Pest/Disease control					
Nat		+0.5	+1	+1	+2.5
Synth	-1	-1	-3	-3	-8
Crop management					
Rotat	+0.5	+0.5	+1		+2
Inter	+1	+1	+1	+1	+4
Weed control					
Herb	-1	-1	-1	-0.5	-3.5
C&C	+1	+1	+1	+1	+4
C. Mgt	+1	+0.5	+1	+0.5	+3

3.2.3 Study area, data collection and analyses methods

Study area

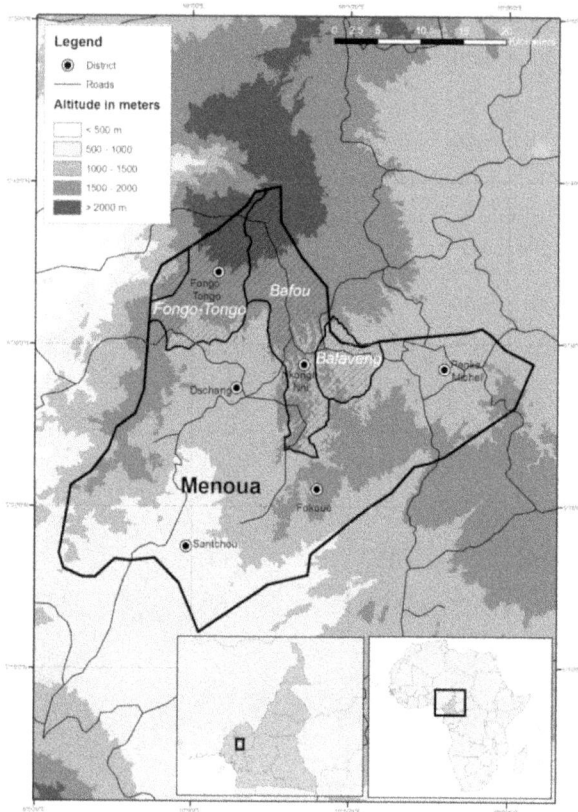

Figure 3.2
Geographical location of research site.

The study was conducted in three villages (Bafou, Baleveng and Fongo-Tongo) found in the WHC (Figure 3.2). The mean annual temperature of the region is estimated at 20°C. The average yearly rainfall is estimated at between 1000 and 2000 mm (Kay*et al.*, 1985), unimodally distributed and the average annual sunshine estimated at 2000 hours. The soils, described as the Djuttitsa soil series (Tchienkoua & Zech, 2003) are derived from trachy-basaltic materials and classified in the USDA soil taxonomy (Soil Survey Staff, 1990) as Andic Palehumult. The farming systems practices and their variations in the studied villages are assumed to be representative of the WHC. While Bafou and Fongo-Tongo villages extend to very high altitudes, Baleveng occupies mostly the low and medium altitude areas. The agricultural system in the WHC is labour intensive; hoes and machetes are the basic farm implements. Livestock comprise an integral part of the farming system, but the progressive conversion of pasture into cropland has caused a reduction in the livestock production of the average household, and so a parallel decline in the amount of manure available for improving soil fertility (Tchienkoua & Zech, 2003). Steep slopes and abundant rainfall are the norms, thus the tasks of field preparation and erosion control are uncommonly difficult for the region's many small holders. Because of high population pressure, land for food crop production is often cultivated on a more or less semi-permanent basis with about one year fallow alternating with about 2–3 years of cultivation. Land preparation often includes partial burning of weeds and residual crop biomass, and ploughing with ridges along the contour (Tchienkoua & Zech, 2003) and sometimes across the contour. It has been estimated that between 250 and 300 kg ha^{-1} year^{-1} of NPK fertilizers (mostly 20:10:10) are applied by the local farmers (Fotsing, 1994).

Data collection

A survey based on interviews with 144 households was carried out in three villages: Bafou, Baleveng and Fongo-Tongo located in the WHC (Figure 3.2) in 2009, 2010 and 2011. The survey questionnaire included questions regarding many current characteristics of the farming system. The questionnaires contained variables on natural conditions, social-economic conditions, infrastructure, structure of agricultural production, inputs for agricultural production, farm output, profitability of agricultural production and farm diversity. The variables analysed in this study were: age of head of household (AHH), size of household (SHH), distance of farthest farm plot (DFF), number of irrigable farm plots (FIRR), number of sole crop species used by the farmer (SCR), number of animal species raised by the household (NAN), number of different farm tools owned by the household (NTO), number of companion crops in intercropping by the farmer (ICR), number of farm plots owned by the household (FOW), distance of the closest farm plot from the homestead (DCF), number of swampy farm plots owned by the household (FSW), average wage paid to hired workers by the farmer (AWA), fallow duration of cropland (FDU) and estimated sustainability score (SUS)

Statistical Analysis

The sustainability parameter was subjected to analysis of variance with the different villages used as the factor and mean separation at 5% probability was carried out using the Student-Newman-Keuls test. For other quantitative variables, means and standard errors were calculated. In the case of analysis of degree and sense of the relationship between qualitative variables, the corresponding contingency tables were constructed and the statistics calculated were used as the basis for the Chi-squared distribution.

Simple correlation analyses were performed for the variables collected from the study area. Groups of correlated variables (excluding sustainability score) were defined for the study site by using factor analysis. Factors were extracted with the factor procedure of the SPSS version 13 package using the principal factor analysis and the Varimax rotation method. New variables were created by standardizing and averaging selected variables from each factor for which the eigenvalues of the correlation matrix were one or greater. The basis for selecting measured variables from each factor is in partial correlation coefficients that are often referred to as factor loadings (Johnson & Wichern, 1992). The new variables are called latent variables. To study the relationships between the latent variables and sustainability score, a multiple regression model was determined for the study area. Sustainability score was the dependent variable and the latent variables were the independent variables. The model was of the form $Y = b_0 + b_1L_1 + b_2L_2 + ... b_nL_n + \varepsilon$, where Y represents estimated sustainability score, b_0 to b_n are coefficients, L_1, to L_n are the latent variables and ε represents residual error.

3.3   Results

3.3.1   Relationship between agricultural production variables and different villages of the WHC

The relationship between the production variables and the different villages are presented in Tables 3.3.

The production variables evaluated were types of farming, labour source, level of education of the heads of households, gender of the heads of households, major means of transportation of the heads of households, major sources of income of the heads of households and the sustainability score of the villages.

The Pearson's chi-square test indicated that the villages were not independent with regard to the labour source and means of transportation. There was a strong relationship (at 5% level of significance) between the different villages and the labour source used for production and between the villages and the means of transportation to their farms. The households of Baleveng and Fongo-Tongo depended more on family members as labour source while the Bafou households used more hired labourers. In the same light, more Bafou households used motorcycles for transportation compared to the Baleveng and Bafou villages. All the other production variables were independent of the villages (Table 3.3). The farming types, level of education, gender of heads of households and major sources of income were similar across the study sites (Table 3.3). With regard to sustainability, the sustainability score was signif-

Table 3.3
Relationship between the villages and production variables.

Variable		Village			X^{2a}
		Bafou	Baleveng	Fongo-Tongo	
Type of farming (%)	N	47	52	45	3.04ns
Mainly crop production		51.1	34.6	37.8	
Crop and livestock production		48.9	65.4	62.2	
Labour source (%)	N	43	49	43	10.73*
Family		23.3	55.1	46.5	
Hired		23.3	12.2	20.9	
Both family and hired		53.5	32.7	32.6	
Level of education of head of household (%)	N	44	51	44	5.32ns
No formal education		4.5	7.8	4.5	
Primary		36.4	49.0	55.3	
Secondary		56.8	43.1	43.2	
Tertiary		2.3	0.0	0.0	
Gender of head of household (%)	N	47	52	45	0.88ns
Male		80.9	75	82.2	
Female		19.1	25	17.8	
Major means of transportation of head of household (%)	N	45	36	47	10.9*
Pedestrian transportation		26.7	58.7	48.6	
Motorcycle		62.2	39.1	43.2	
Motor vehicle		11.1	2.2	8.1	
Major sources of income by household (%)	N	47	52	42	0.12ns
Farming		27.7	30.8	28.6	
Farming and off-farm activities		72.3	69.2	71.4	
Sustainability score+		0.26a	0.44b	0.31a	

+Means followed by the same letter are not significantly different (p<0.05)
a*indicates significant difference at 5% probability level while ns indicates non-significant difference at 5% probability level for the different Chi square contingency table analysis.

icantly higher (p < 0.05) in Baleveng village than in Bafou and Fongo-Tongo villages (Table 3.3).

3.3.2 Determinants used for the factor analysis.

The mean values and the variability of the determinants selected for factor analysis are presented in Table 3.4.

Table 3.4
Descriptive statistics for the selected variables used for factor analysis in the study.

Variable	Mean	Maximum	Minimum	Standard deviation
AHH (years)	51.65	91	23	13.34
SHH	11.19	35	2	6.54
FOW	3.88	1	20	2.81
DFF (km)	4.39	0	35	5.64
FIRR	0.90	0	10	1.44
FSW	0.56	0	6	1.06
SOLE	1.53	0	7	1.75
NAN	1.74	0	4	1.13
NTOO	3.27	1	7	1.16
NINT	3.19	0	10	2.24
SUS	0.34	0.09	0.94	0.18

Some of the variables collected in the study area were very closely related and had to be eliminated after preliminary analysis.
The correlation coefficient of all pairs of the variables is shown in Table 3.5.

The highest number of correlations were recorded with the number of farm plots owned by the household (FOW) followed by the sustainability score (SUS) and the number of companion crops used in intercropping (NINT) while the least number of correlations was between the distance of the furthest farm plot from the homestead (DFF) and the other variables. The number of significant correlations in Table 3.5 suggested that a multivariate approach to data reduction was productive.

Table 3.5
Simple correlation coefficients for the variables studied in the study area

Variable	AHH	SHH	FOW	DFF	FIRR	FSW	SOLE	NAN	NTOO	NINT
AHH										
SHH	0.26**									
FOW	-0.27**	0.21*								
DFF	-0.08	-0.05	-0.01							
FIRR	-0.37**	0.03	0.32**	-0.03						
FSW	-0.01	0.06	0.33**	-0.10	0.36**					
SOLE	-0.27**	-0.04	0.13	0.04	0.35**	0.20*				
NAN	-0.09	0.21*	0.22*	0.13	0.05	0.13	0.02			
NTOO	0.14	0.17	0.10	-1.11	0.09	0.12	0.19*	0.13		
NINT	-0.05	0.20*	0.19*	-0.25**	0.06	-0.04	0.14	0.19*	0.36**	
SUS	0.28**	0.07	-0.27**	-0.07	-0.39**	-0.24**	-0.23**	-0.08	-0.17	-0.22*

*and ** represent significance at 5% and 1% probability levels respectively.

The factor analysis extracted four factors (Table 3.6) from 10 explanatory variables initially identified (see Table 3.4) with eigenvalues greater than one.

Table 3.6
Results of principal components factor analysis and varimax rotation of the first three factors.

Variable	Factor 1	Factor 2	Factor 3	Factor 4
AHH	0.09	-0.05	-0.83	-0.06
SHH	0.37	0.27	-0.45	0.32
FOW	0.18	0.59	0.21	0.35
DFF	-0.39	-0.20	0.20	0.70
FIRR	0.06	0.61	0.50	0.01
FSW	-0.06	0.86	-0.04	-0.08
SOLE	0.26	0.22	0.60	-0.05
NAN	0.25	0.18	-0.17	0.71
NTOO	0.71	0.10	-0.03	0.00
NINT	0.84	-0.08	0.14	0.05
Eigen value	1.68	1.67	1.64	1.23
% variance	16.79	16.67	16.42	12.67
% Cum. Var.	16.79	33.46	49.88	62.15

These four factors explained 62.15% of the total variance. The signs of the factor loadings provide information on how these variables relate when representing the common factor. It is observed that the most important variable in the first component is the number of companion crops used in intercropping (NINT) and its influence is positive in the component. The other important variable which exhibits a positive influence on the first component is the number of farm tools owned by the household (NTOO). The second component is positively influenced by the number of farm plots in swampy areas owned by the household (FSW), the number of total farm plots owned by the household (FOW) and the number of irrigable farm plots owned by the household (FIRR). The third component is negatively influenced both by the age of the head of the household (AHH) and the size of the household (SHH) and positively influenced by the number of crops produced under sole cropping (SOLE). The fourth factor is positively influence by the distance of the furthest farm plot from the homestead and the number of animal species raised by the household.

3.3.3 Relative importance of the factors influencing the sustainability of the farming system

Table 3.7 shows coefficients and statistics of models relating sustainability scores with the latent variables for the study area.

Table 3.7
Coefficients and statistics of multiple regression models relating sustainability with the latent variables identified for the three villages.

Intercept	Factor 1	Factor 2	Factor 3	Factor 4	R2	P>F
0.32 (0.02)[a]	-0.03 (0.02)	-0.04 (0.02)	-0.06 (0.02)	-0.02 (0.02)	0.23	0.001

[a]Numbers in parenthesis are standard errors of the estimates.

All the latent variables had negative coefficients indicating that they are all negatively related to the sustainability score.

3.3.4 Main constraints influencing agricultural production in the study area

The constraints faced by the farmers are summarized on Table 3.8.

Table 3.8
Percentage of farmers' priority of agronomic production constraints

Constraint	Bafou	Baleveng	Fongo-Tongo	χ2
Poor yield	71.1	75.5	77.8	Ns
Poor road infrastructure	55.6	36.7	68.9	**
Problems with crop pests	60.0	64.6	64.4	Ns
High cost of inputs	82.2	79.6	73.3	Ns
Low price of outputs	62.2	38.8	57.8	Ns

**significant at $p < 0.01$, ns not significant

Except for their views on road infrastructure, all the other constraints were similar for farmers in all the villages.

3.4   Discussions

3.4.1 Relationship between agricultural production variables and the different villages

Households in all the villages studied showed a similar interest in both crop and livestock production. Livestock production was valued as the second most important activity after crop production. Animals were used mainly for social events such as payment of dowry, guests' reception and ceremonies (wedding, funerals etc.). Cash crops produced were vegetable crops concentrated around the higher altitudes and irrigable lands. The most important of these crops were potatoes, cabbages, carrots, leeks, beetroots, onions, and tomatoes which were cultivated by sole cropping. The vegetable crops were principally for the market. These crops consumed huge amounts of off-farm chemical inputs (fertilizers, insecticides, fungicides, and herbicides) and employed a significant number of wage labourers for all the production stages (from land preparation to harvesting and transportation). Farmers produced an average of more than two crops a year especially where irrigation was feasible. Food crop production was concentrated at the lower altitudes and was principally produced by intercropping. The major crops found in the multiple cropping fields included maize (usually in the first season), beans, aroids, plantains, sweet potatoes, groundnut, and assorted leafy vegetables.

More wage labourers were needed in the high input vegetable production system common in the area which explained the importance of this practice in the three villages studied. Baleveng was least dependent on off-farm chemical inputs and hence the lower number of wage labourers employed. This could be due to the fact that high altitudes favourable for cool-season crops that are highly dependent on chemical inputs are not available in Baleveng. As a result farmers in Baleveng were not greatly involved in cool-season vegetable crops. Labour type and requirement in the study area depended on household size; number of farms owned and the size of adult household members. Labour is the primary instrument for increasing production within the framework of traditional agriculture. Amaza et al. (2009) found that fertilizer and hired labour were the major factors associated with changes in the output of food crops in the Borno state, Nigeria. In their study they found that chemical fertilizer and hired labour had significant positive effects on output. This is similar to the results of this study as the use of off-farm inputs and hired labour explain the differences among the villages studied in the same manner.

The fact that the head of many households had at least finished high school was an encouraging sign as it implied they were forward-looking and open to the idea of change. Many studies have revealed that the level of education is a factor in helping farmers to use production information efficiently (Hayami, 1969; Lockheed et al., 1980; Phillips, 1994; Wang et al., 1996; Yang, 1997). Education also influential when it comes to making use of opportunities available to improve livelihood strategies, enhance food security, and reduce the level of poverty. It affects the level of exposure to new ideas and managerial capacity in production as well as the perception of the household members on how to adopt and integrate innovations into the household's survival strategies.

Females who headed households in our research site were mostly widows. Female headship is often believed to increase the likelihood of the household being poor but World Bank data indicated that while this may be true in Asia and Latin America it is less obvious in Africa (Chant, 2003). In the WHC, many women compete well with men in nearly all activities. However the rural-urban division of labour has required women to undertake all the agricultural tasks, thus curtailing the extent to which they can participate in the labour market as also noted by Gwaunza (1998).

The Bafou farmers had an edge over the other villages in their possession of a greater number of motorbikes which facilitated their farming operations. Roads and transportation facilities are essential for the sustainability of agricultural production in Sub-Saharan Africa as it positively impacts factors such as mobility (John & Carapetis, 1991), the adoption of high yielding varieties, high productivity crops and bigger farm size (Sieber, 1999). The greater involvement and higher success in agriculture of the Bafou farmers could thus be attributed to the better transportation facilities they possessed.

The low percentage of farmers whose main occupation was agricultural is characteristic of the high number of pluriactive households in the WHC. The dependence of farm families on off-farm activities as an income source is absolutely necessary, owing to the uncertainties commonly dictated by weather, market prices and attacks by pests. In sub-Saharan Africa, it is common for some farm household members to engage in other non-farming occupations to complement their earnings from farming. A study by Herbert (1996) in Burundi reveals the need of income diversification through extra-agricultural activities to complement farming.

On the whole the sustainability scores of all the sites in our study were below average due to their reliance on off-farm non-renewable inputs and the reduction of the genetic base through sole cropping. Intensive agricultural practices carried out by most farmers in the WHC necessitates transporting the inputs which it uses from more distant sources, deriving an increasing proportion of its energy supplies from non-renewable sources, depending upon a narrower genetic base and having an increasing impact on the environment. This is particularly reflected in its heavy reliance on chemical fertilizers, and pesticides, exploitation of vulnerable lands in the cultivation of the vegetable cash crops in the region, all of which contribute to environmental pollution, habitat destruction and risks to human health and welfare as observed by Hodge (1993). Amongst the three villages studied, a significantly higher sustainability score was recorded by the Baleveng village owing to the absence of high altitude areas favourable to the intensive off-farm dependent crop production system common in the Bafou and Fongo-Tongo villages. As such, farmers in Baleveng village relied less on off-farm inputs because a majority of the farmers practiced intercropping where annidation or complementarity provided appropriate growing conditions (Trenbath, 1976). Intercropping has been an agricultural practice for thousands of years (Kass, 1978), which testifies to its level of sustainability. Ofori and Stern (1987) suggested that intercropping was more efficient than mono-cropping in the exploitation of limited resources. Food challenges will be met using environmentally friendly and socially equitable technologies and methods, in a world with a shrinking arable land base (which is also being diverted to produce biofuels), with less and more expensive petroleum, increasingly limited supplies of water and nitrogen, and within a scenario of a rapidly changing climate, social unrest, and economic uncertainty (IAASTD, 2009). The only agricultural system that will be able to confront future challenges is one that will exhibit high levels of diversity, productivity, and efficiency.

3.4.2   Determinants associated with the farming systems of the WHC

There were various degrees of correlation among the determinants associated with the farming systems of the WHC. The sustainability score in the whole re-search area was negatively related to the number of farms owned by households (FOW), the number of irrigable plots owned by households (FIRR), the number of farm plots in swampy areas (FSW), the number of crops used in sole crop-ping (SOLE), the number of companion crops used in intercropping (NINT) and positively related to the age of the head of household (AHH). This indicated that the older heads of households carried out more sustainable farm practices with respect to seed source, soil fertility, crop management, pest and disease control and weed control. Household characteristics are thus important deter-minants of the farming system. However, Rougoor *et al.* (1998) found that the influence of the age of the head of the household on farm productivity was very diverse. Other studies found that age had a positive effect on productivity (Kali-rajan & Shand, 1986; Stefanou & Sexena, 1988) while Adubi (1992) revealed that age, in correlation with farming experience, had a significant influence on the decision making process of farmers with respect to risk aversion, adoption of improved agricultural technologies, and other production-related decisions. Age has been found to determine how active and productive the head of the household would be. It has also been found to affect the rate at which house-holds adopted innovations, which in turn, affects household productivity and livelihood improvement strategies (Dercon & Krishnan, 1996). All the determi-nants that had negative relations with the sustainability score are linked directly or indirectly to either dependence on off-farm agrochemical inputs or soil min-ing. Increasing numbers of companion crops (NINT) leads to less sustainability because of the intensity of land use over space by the high density of species with varied requirements as also noted by Fasching (2001).

3.4.3   Influence of the determinants on sustainability

After factoring the correlation matrix by the principal component method, the first four factors explained 62.15 % variation. The first latent variable had high loadings with the number of farm tools used by the household (NTOO, 0.71) and the number of companion crops used in intercropping (NINT, 0.84). This means that both NTOO and NINT lie near the first axis. The first axis was termed *land use intensity over space* because it is most correlated with compo-nents that have to do with land use in space. The second axis was most corre-lated with practices that require high off-farm inputs for intensive production (FOW, FIRR, FSW); overall this axis appears to measure *the intensity of off-farm inputs*. The third axis (*household adjustment factor*) was most correlated with components that influence the household (AHH, SHH) lying near the third axis, and the number of sole crops used by the household (SOLE) lying on the opposite end of the third axis. The fourth axis was most correlated with components that have to do with the movement of the household(DFF) and the number of animals produced by the household (NAN); overall this factor appears to measure the *mobility of the household* owing to the fact that animal production is not very intensive in the area.

All the latent variables had negative correlations with the sustainability score. The negative sign for the land use intensity characterised by "plant biodiversity" (NINT) seems agronomically unreasonable as it should not decrease sustainability since it implements many different functions such as biomass decomposition, nutrient cycling, soil structure enhancement, pest regulation, pollination, detoxification, local hydrological process regulation and macro-climate control (Altieri, 1999). Having less diversity than needed can eventually lead to production and profitability problems. Adding more diversity than needed can reduce efficiency since it increases the number of crops that must be managed, handled, and marketed (Fasching, 2001). This explains why increasing the number of companion crops in intercropping will decrease sustainability. Though intercropping is envisaged as a contributor to sustainability, human efforts are required to make this happen. The suggested advantages of the intercropping system include yield stability under adverse environmental conditions, efficient use of limited growth resources, biological diversity, and potential control of pests and diseases. Many studies have shown that intercropping systems out yielded sole cropping systems of component crops (Baumann *et al.*, 2001; Lesoing & Francis, 1999; Ghaffarzadeh *et al.*, 1997; Fortin*et al.*, 1994; Mandal *et al.*, 1990).

The negative relationship between the intensity of off-farm inputs and sustainability should be obvious. Intensification and concomitant increased use of inputs in agricultural production has led to environmental pollution and low quality products (Rahman & Thapa, 1999). In order to combat this,efforts are now required to minimise off-farm inputs in order to guarantee the sustainability of farming systems. Sustainable agriculture is often viewed as low input and regenerative (Lockeretz, 1989; Reijntjes *et al.*, 1992), making better use of the farm's internal resources through the incorporation of natural processes into agricultural production and the greater use of knowledge and skills of farmers to improve their self-reliance and capacities.

The household adjustment factor had a negative relationship with sustainability. This latent variable was characterised by a negative sign for both the age of members of the household and size of the household. Taking this into account, it would mean that these components of the household adjustment factor have positive relationships with the sustainability score given that the product of two negatives is positive. The findings of this study thus suggest that sustainable farming practices in the research area are executed more by more populated households and households headed by older people. Many studies have shown the positive correlation between age and environmentally friendly agricultural practices. In Mexico, age was found to play a significant role in determining how much diversity farmers maintain. Almost 50% of the farmers growing significant numbers of traditional cultivars were over 56 years (Morales & Quinones, 2000). Wakeyo and Gardebroek (2013) postulated that in developing countries, households allocate financial resources to buying inputs after putting aside a minimum amount for household food, especially when there is a credit constraint. As such, some households exhaustively consume their harvest and are later constrained to buy inputs such as fertilizer. This attitude is positively related to the size of the household which explains the dependence on natural resources for farming by more populated household and hence the positive relationship between sustainability and the size of the household shown by the results of this study.

With respect to the mobility component, the results of this study suggest that farmers whose farm plots are furthest from their homestead carry out less sustainable practices. This can be justified by the fact that suitable farming areas for the important cash crops of the area are located at high altitudes which are further from the homestead. The method of production of these cash crops require intensive use of agrochemical and improved seeds all of which are negative contributors to sustainability based on our assumptions. Generally crop diversity decreases with the distance of the farm plots from the homestead. In Ethiopia, Deribe (2000) showed that sorghum diversity was related to distance from the homestead: the nearer the plot to the homestead, the larger the number of varieties grown. The use of locally adapted cultivars is usually associated with limited chemical inputs and these can also serve to maintain ecosystem health and improve soil structure (Vandermeer, 1995; Wood & Lenné, 1999). Cultivars adapted to particular micro-niches are often one of the few resources available to resource-poor farmers to maintain or increase production from their fields (Jarvis *et al.*, 2000).

3.4.4 Main constraints that influence agricultural production of the area

All farmers make decisions in a complex environment in which broad contextual factors, such as markets, public policies (including regulation), and social institutions, create opportunities but also create barriers to change. With regard to farmers' priorities, households in the Baleveng village did not consider the road infrastructure as a problem simply because the tarmac road connecting the division and the region cuts across the centre of the village. This makes it fairly easy for inhabitants to move compared to other villages where the transportation of people and goods is sometimes impracticable especially in the rainy season. The problems related to poor yields and crop pests may be linked to a lack of information on improved technology. If agricultural productivity is to grow in Africa, research and extension services need to develop and disseminate science-based information about improved technologies that address the resource constraints and risks faced by the majority of Africa's farmers (Snapp et al., 2003). Agricultural advisers are few and far between in the WHC which explains many of the farmer's problems. There is a need to upgrade the researcher-agricultural adviser-farmer network in the WHC. Limited adoption of recommended technologies must be expected if there is a poor connection between research, technical advisers and African smallholders (Meertens, 2003). Cameroon and many other countries have removed subsidies on fertilizers since the collapse of the coffee and cocoa markets making their affordability extremely difficult for small scale farmers. Integration of crops and livestock can lead to more efficient use of land unsuitable for crop production. It can provide a use for crop residues and by-products, provide manure, and provide a source of income, savings, and investment.

The significance of household size in farming hinges on the fact that the availability of labour for farm production, the total area cultivated for different crop enterprises, the amount of farm produce retained for domestic consumption and the marketable surplus, are all determined by the size of the farm household (Amaza et al., 2009). Increasing dependence on hired labour in our study site was due to the decreasing size of households influenced by rural-to-urban migration.

3.5   Conclusion

With respect to the different variables that determine agricultural production, the results of this study show that each of the different villages studied, have much in commom can benefit equally from the same improved technologies and recommendations. Research needs to address land use intensity, off-farm inputs intensity, household adjustment factors and the mobility of the household. Common features of the farming systems in WHC are that they manage natural and economic resources and conditions that vary in time, with limited production alternatives while facing relatively low profit. Both the variation in farming systems and the common characteristics of farms lead to uncertainties about the effectiveness of decisions, from a farmer's and from a policy perspective. In spite of rapid social change, traditional hierarchical structures still influence village life so village leaders should be involved in the introduction of agricultural change. Agriculture that is truly sustainable will not mean business as usual. It will be a type of agriculture that will provide environmental, economic and social opportunities for the benefit of present and future generations, while maintaining and enhancing the quality of the resources that support agricultural production. This will not place the emphasis on maximizing yields and economic returns, but will rather focus on optimizing productivity and conserving the natural resource base. Well designed intercropping systems and the use of natural organic resources as sources of plant nutrients would benefit and satisfy both the producers and consumers in the system. Intercropping is the intensification of land and resource use in the space dimension. This can lead to: enhanced efficiency of incident light use with two or more species that can occupy the same land area and have different patterns of foliage display; different rooting patterns can explore a greater soil volume with roots of different depths; competition with weeds from a combination of species occupying two or more niches in the cropping environment can effectively reduce weed germination; a mixture of crops can provide a buffer against losses to plant diseases. In order to fill in the gaps in our understanding of the effects of emerging farming systems on sustainability in WHC, there is a real need for system-level (holistic-whole-farm approach) studies for a more detailed picture of the situation.

Acknowledgements

This research was made possible by funding from Volkswagen foundation. The authors wish to thank Maarten van't Zelfde of the Institute for Environmental Sciences (CML), Leiden, for producing the map of the research site. We are very grateful to the extension workers of the Menoua Division who participated in the data collection and the household members who provided the useful data for the study.

References

Adubi, A.A. (1992) *An empirical analysis of production risks and attitudes of small farmers in Oyo State, Nigeria*. PhD thesis, University of Ibadan, Nigeria.

Ahmed, M.M., Rohrbach, D.D., Gono, L.T., Mazhangara, E.P., Mugwira, L., Masendeke, D.D. & Alibaba, S. (1997) *Soil fertility management in communal areas of Zimbabwe: current practices, constraintsand opportunities for change. Results of a diagnostic survey*. Southern Eastern Africa Region working paper no. 6. Bulawayo, Zimbabwe: International Crops Research Institute for the Semi-Arid Tropics (ICRISAT).

Altieri, M.A. (1995) *Agroecology: The scientific basis of alternative agriculture*. Boulder, CO: Westview Press.

Altieri, M.A. (2000) Agroecology: the science of natural resource management for poor farmers in marginal environments. *Agriculture, Ecosystems and Environment* 93: 1-24.

Altieri, M.A., Francis, C.A., Schoonhoven, A. van & Doll, J.D. (1978) A review of insect prevalence in maize (*Zea mays* L.) and bean (*Phaseolus vulgaris* L.) polycultural systems. *Field Crops. Res.* 1: 33-49.

Amaza, P.S., Bila, Y., & Iheanacho, A.C. (2006) Identification of Factors that Influence Technical Efficiency ofFood Crop Production in West Africa: Empirical Evidence fromBorno State, Nigeria. *Journal of Agriculture and Rural Development in the Tropics and Subtropics* 107(2): 139-147.

Amaza, P.S, Abdoulaye, T., Kwaghe, P. & Tegbaru, A. (2009) *Changes in household food security and poverty status in PROSAB areas of southern Borno State, Nigeria.* Promoting Sustainable Agriculture in Borno State (PROSAB). Ibadan, Nigeria: International Institute of Tropical Agriculture, 40 p.

ATTRA (2003) Applying the principles of sustainable farming. *Appropriate Technology Transfer for Rural Areas.* www.attra.ncat.org, accessed in May 2009.

Badgley, C., Moghtader, J., Quintero, E., Zakem, E., Chappell, M.J., Aviles-Vazquez, K., Samulon, A., Perfecto, I. (2007) Organic agriculture and the global food supply. *Renew. Agric. Food Syst.* 22: 86-108.

Baijukya, F.P. & Steenhuijsen, P.B. de (1998) Nutrient balances and their consequences in the banana based land use systems of Bukoba district, North West Tanzania. *Agriculture, Ecosystems and Environment* 71: 147-158.

Bartlett, B., Blake, B. & McCarl, B. (1982) Goal programming via multidimensional scaling applied to Senegalese subsistence farms. *American Journal of Agricultural Economics* 64: 720-727.

Baumann, D.T., Bastiaans, L. & Kropff, M.J. (2001) Competition and crop performance in a leek-celery intercropping system. *Crop Sci.* 41: 764-774.

Bergeret, P. & Djoukeng, V. (1993) Evaluation économique des systèmes des cultures en pays Bamiléké (Ouest Cameroun). *Cahiers Agricultures* 2: 187-196.

Beus, C.E. & Dunlop, R.E. (1994)Agricultural paradigms and the practice of agriculture. Rural Sociology 59(4): 620-635.

Binswanger, H.P. (1980) Attitudes towards risk: experimental measurement in rural India. *American Journal of Agricultural Economics* 62: 395-407.

Boserup, E. (1965) *The conditions of agricultural growth.* London: Earthscan.

Bull, D. (1982) *A Growing Problem. Pesticides and the Third World Poor.* Oxford: OXFAM, 192 p.

Chant, S. (2003) *New Contributions to the Analysis of Poverty: Methodological and Conceptual Challenges to Understanding Poverty from a Gender Perspective, CEPAL, U.N.* Santiago, Chile: Women's Development Unit, .

Crosson, P.R. (1992) Sustainable agriculture. *Quarterly Newsletter, Resources Future* 106: 14-17.

Dalsgaard, J.P.T., Lightfoot, C. & Christensen, V. (1995) Towards quantification of ecological sustainability in farming systems analysis. *Ecological Engineering* 4: 181-189.

De Jager, A., Kariuki, I., Matiri, F.M., Odendo, M. & Wanyama, J.M. (1998) Monitoring nutrient flows and economic performance in African farming systems (NUT-

MON). IV. Linking farm economic performance and nutrient balances in three districts in Kenya. *Agriculture, Ecosystems and Environment* 71: 81-92.

Dercon, S. & Krishnan, P. (1996) Income portfolios in rural Ethiopia and Tanzania: choices and constraints. *Journal of Development Studies* 32(6): 850-875.

Deribe, S. (2000) *A study on the agrobiodiversity with special emphasis on the relationship between distance from homestead and crop diversity on-farm in Fontanina South Welo.* MSc Thesis University of Addis Ababa.

DFID (2002) *Sustainable Agriculture.* Resource Management Keysheet 10. London, UK: DFID.

Dumanski, J. & Pieri, C. (1996) Application of the pressure-state-response framework for the land quality indicators (LQI) program. In: *Land quality indicators and their use in sustainable agriculture and rural development* (p. 41). Proceedings of the workshop organized by the Land and Water Development Division FAO Agriculture Department, January 25-26.

Ewel, J.J. (1999) Natural systems as a model for the design of sustainable systems of land use. *Agrofor. Syst.* 45(1/3): 1-21.

FAO (2002) *Food and Agricultural Organization.* Rome: FAOSTAT data base.

Fasching, R.A. (2001) *Crop rotation intensity rating and diversity index. United States Department of Agriculture Natural Resources Conservation Service.* Crop Production Agronomy Technical Note, No. 150.14. Flores, M. (1989) Velvetbeans: an alternative to improve small farmers' agriculture. *ILEIA Newsletter* 5: 8-9.

Floret, C. (1998) *Raccourcissement du temps de jachère, biodiversité et développement durable en Afrique Centrale (Cameroun) et en Afrique de l'Ouest (Mali, Sénégal).* Final Report, Comm. des Communautés européennes, Contrat TS3-CT93-0220 (DG 12 HSMU). Paris: IRD, 245 p.

Fortin, M.C., Culley, J. & Edwards, M. (1994) Soil water, plant growth, and yield of strip intercropped corn. *J. Prod. Agric.* 7: 63-69.

Fotsing J.M. (1994) L'évolution du bocage bamiléké: exemple d'adaptation traditionnelle à une forte démographie, in: introduction à la gestion conservatoire de l'eau, de la biomasse et de la fertilité des sols (GCES). *Bull. Pédologique de la FAO* 70: pp. 292-307.

Funes-Monzote, F.R. (2009) *Agricultura con futuro. La alternativa agroecológica para Cuba.* Estación Experimental Indio Hatuey, Universidad de Matanzas.

Gagné, W.C. (1982) The transformation and intensification of shifting agriculture: past and present conservation practices. In: Moranta, L., Pernetta, J. & Heaney, W. (eds) *Traditional Conservation in Papua, New Guinea: Implications for Today.* Inst. Appl. Soc. Econ. Res. Monogr. 16, pp. 153-58.

Ghaffarzadeh, M., Garcia Prechac, F. & Cruse, R.M. (1997) Tillage effect on soil water content and corn yield in a strip intercropping system. *Agron. J.* 89: 893-899.

Girardin, P., Bockstaller, C. & Werf, H.M.G. van der (1999) Indicators: tools to evaluate the environmental impacts of farming systems. *Journal of Sustainable Agriculture* 13: 5-21.

Gomez, A.A., Kelly, D.E., Syers, J.K. & Coughlan, K.J. (1996) Measuring Sustainability of Agricultural Systems at the Farm Level. *Methods for Assessing Soil Quality, SSSA Special Publication* 49: 401-409.

Gras, R., Benoit, M., Deffontaines, J.P., Duru, M., Lafarge, M., Langlet, A. & Osty, P.L. (1989) *Le fait technique en agronomie. Activité Agricole, Concepts et Méthodes d'Etude*. Paris, France: Institut National de la Recherche Agronomique, L'Hamarttan, 184 p.

Gwaunza, E. (1998) The impact of labour migration on family organisation in Zimbabwe. In: Sachikonye, L. (ed.) *Labour Markets and Migration Policy in Southern Africa*. Harare: Sapes Books, pp. 49-55.

Hansen, J.W., Knapp, E.B & Jones, J.W. (1997) Determinants of sustainability of a Colombian hillside farm. *Expl Agric.* 33: 425-448.

Hayami, Y. (1969) Sources of agricultural productivity gap among selected countries. *American Journal of Agricultural Economics* 51(2): 564-575.

Herbert, J.P. (1996) A study of the diversification of farm activities resulting from demographic and land pressure; a case study of Burundi. *Tropiculture* 14(1): 17-23.

Hodge, I. (1993) *Sustainability: Putting Principles into Practice. An Application to Agricultural Systems*. Paper presented to 'Rural Economy and Society Study Group, Royal Holloway College, December 1993.

Hoffman, I., Gerling, D., Kyogwom, U.B. & Mane-Bielfeldt (2001) Farmers' management strategies to maintain soil fertility in a remote area in northwestNigeria. *Agriculture, Ecosystems and Environment* 86(3): 263-275.

Howlett, D.J.B. (1996) Development of social, economic and biophysical indicators for sustainable land management in the South Pacific. In: Howlett, D. (ed.) *Sustainable Land Management in the South Pacific*. Network Document No 19. International Board for Soil Research and Management, Bangkok, Thailand: IBSRAM.

IAASTD (International Assessment of Agricultural Knowledge, Science and Technology for Development) (2009) Agriculture at a Crossroads. In: *International Assessment of Agricultural Knowledge, Science and Technology for Development Global Report*. Washington, DC: Island Press.

IFAD (2011) *The world's population is about to hit 7 billion*. International Fund for Agricultural Development.

Ikerd, J. (1993) Two related but distinctly different concepts: organic farming and sustainable agriculture. *Small Farm Today* 10(1): 30-31.

IRRI International Rice Research Institute (1976) *The IRRI Reporter*, April. Los Banos, Philippines, 4 p.

IRRI (International Rice Research Institute) (2000) *Final Report of the project: Safeguarding and preservation of the biodiversity of the rice gene pool*.

Isaac, M.E., Dawoe, E. & Sieciechowicz, K. (2009) Assesing local knowledge use in agroforestry management with cognitive maps. *Environmental Management* 43: 1321-1329.

Jarvis, D. & Hodgkin, T. (2000) Farmer decision making and genetic diversity: linking multidisciplinary research to implementation on-farm. In: Brush, S.B. (ed.) *Genes in the Field: On-Farm Conservation of Crop Diversity*. Boca Raton, FL, USA: Lewis Publishers, pp. 261-278.

John, D.N.R. & Carapetis, S. (1991) *Intermediate Means of Transport in Sub-Saharan Africa: Its Potential for Improving Rural Travel and Transport*. World Bank Technical Paper No. 161. Africa Technical Department Series. Washington, D.C., USA: World Bank.

Johnson, R.A. & Wichern, D.W. (1992) *Applied multivariate analysis*, 3d edition, 642 pp. Englewood Cliffs, NJ: Prentice Hall Inc.

Juo, A.S.R. & Manu, A. (1996) Chemical dynamics in slash-and-burn agriculture. Agriculture. *Ecosystems and Environment* 58: 49-60.

Kalirajan, K. & Shand, R.T. (1986) Estimating Location-Specific and Firm-Specific Technical Efficiency: An Analysis of Malaysian Agriculture. *Journal of Economic Development* 11: 147-160.

Kass, D.C.L. (1978) Poly culture cropping systems: review and analysis. *Cornell Int. Agric. Bull.* No. 32. Ithaca, NY: Cornell Univ., 69 p.

Kay, M., Stephens, W. & Carr, M.K.V. (1985) Prospects for small-scale irrigation in sub-Saharan Africa. *Outlook on agriculture* 14(3): 115-121.

Kotto-Same, J., Woomer, P.L., Moukam, A. & Zapfack, L. (1997) Carbon dynamics in slash-and-burn agriculture, and land-use alternatives of the humid forest zone in Cameroon. *Agriculture, Ecosystems and Environment* 65: 245-256.

Lesoing, G.W. & Francis, C.A. (1999) Strip intercropping effects on yield and yield components of corn, grain sorghum, and soybean. *Agron. J.* 91: 807-813.

Lockheed, M.E., Jamison, D.T. & Lau, I.J. (1980) Farmer education and farm efficiency: a survey. *Economic Development and Cultural Change* 29: 37-76.

Lockeretz, W. (1989) Problems of evaluating the economics of ecological agriculture. *Agriculture, Ecosystems & Environment* 27: 67-75.

Lynam, J.K. (1994) Sustainable growth in agricultural production: the links between production, resources and research. In: Goldsworthy, P. & Penning de Vries, F.W.T. (eds) *Opportunities, Use and Transfer of Systems Research Methods in Agriculture to Developing Countries*. Dordrecht: Kluwer Academic Publishers, pp. 3-28.

Malthus, T.R. (1989) *An essay on the principle of population*. Cambridge: Cambridge University press.

Mandal, B.K., Dhara, M.C., Mandal, B.B., Das, S.K. & Nandy R. (1990) Rice, mungbean, soybean, peanut, ricebean, and blackgram yields under different intercropping systems. *Agron. J.* 82: 1063-1066.

Mead, R., Curnow, R.N. & Hasted, A.M. (2003) *Statistical methods in Agriculture and Experimental Biology*, 3rd Edition, Washington, USA: Chapman & Hall/CRC, 403 p.

Mead, R. & Willey, R.W. (1980) The concept of a 'land equivalent ratio' and advantages in yield from intercropping. *Exp. Agric.* 16: 217-218.

Meertens, H.C.C. (2003) The prospects for integrated nutrient management for sustainable rainfed lowland rice production in Sukumaland, Tanzania. *Nutrient Cyclingin Agroecosystems* 65(2): 163-171.

Mitchell, G., May, A. & McDonald, A. (1995) PICABUE: a methodological framework for the development of indicators of sustainable development. *Int. J. Sustain. Dev. World Ecol.* 2: 104-123.

Morales-Valderrama, C. & Quiñones-Vega, T. (2000) Social, cultural and economic data collection and analysis including gender: methods used for increasing access, participation and decision-making. In: Jarvis, D., Sthapit, B. and Sears, L (eds) *Conserving agricultural biodiversity in situ*, pp. 49-50.

Müller, S. (1996) *How to measure sustainability: an approach for agriculture and natural resources*. Discussion Paper series on sustainable agriculture and natural resources. IICA/BMZ/GTZ.

Müller, S. (1998) *Evaluating the sustainability of agriculture*. Eschborn, Germany: GTZ.

OECD (Organisation for Economic Co-operation and Development) (2001) *Environmental indicators for agriculture: Concepts and framework*, Volume 3. Paris: OECD.

Oenama, O., Kros, H. & de Vries, W. (2003) Approaches and uncertainties in nutrient budgets: implications for nutrient management and environmental policies. *Europ. J. Agronomy* 20: 3-16.

Ofori, C.F. & Stern, W.R. (1987) Cereallegume intercropping systems. *Adv. Agron.* 26: 177-204.

Oreszczyn, S., Lane, A. & Carr, S. (2010) The role of networks of practice and webs of influencers on farmers' engagement with and learning about agricultural innovations. *J. Rural Studies* 26: 404-417.

Oyatoye, E.T.O. (1994) The impact of Rural Roads on Agricultural Development in Nigeria: A case study of Kwara State. *Ife Journal of Agriculture* 16: 114-122.

Parrot, N. & Marsden, T. (2002) *The real Green Revolution: Organic and Agroecological farming*. IntheLondon: GreenpeaceEnvironment Trust, pp. 1-6.

Pham, J.L., Quilloy, S., Huong, L.D., Tuyen, T.V., Minh, T.V. & Morin, S. (1999) *Molecular diversity of rice varieties in Central Vietnam*. Paper presented at the Workshop of the participants of the project "Safeguarding and Preserving the Biodiversity of the Rice Genepool. Component II: On-farm Conservation. International Rice Research Institute, Los Baños, Philippines, May 17-22.

Phillips, J.M. (1994) Farmer education and farmer efficiency: a meta-analysis. *Economic Development and Cultural Change* 43: 1439-1465.

Place, F., Barrett, C.B., Freeman, H.A., Ramisch, J.J., Vanlauwe, B. (2003) Prospects for integrated soil fertility management using organic and inorganic inputs: evidence from smallholder African agricultural systems. *Food Policy* 28: 365-378.

Pretty, J.N. (1995) *Regenerating agriculture: Policies and practice for sustainability and self-reliance*. London: Earthscan Publications Limited.

Pretty, J. & Hine, R. (2001) *Reducing food poverty with sustainable agriculture: a summary of new evidence*. UK: University of Essex Centre for Environment and Society.

Pretty, J.N., Noble, A.D., Bossio, D., Dixon, J., Hine, R.E., Penning de Vries, F.W.T. & Morison, J.I.L. (2006) Resource-conserving agriculture increases yields in developing countries. *Environmental Science and Technology (Policy Analysis)* 40(4): 1114-1119.

Quisumbing, A., Lawrence, H. & Pena, C. (2001) Are women over-represented among the poor? An analysisof poverty in 10 developing countries. *Journal of Development Economics* (66): 225-269.

Rahman, S. & Thapa, G.B. (1999) Environmental impacts of technological change in Bangladesh agriculture: farmers' perceptions and empirical evidence. *Outlook on Agriculture* 28(4): 233-238.

Rasul, G. & Thapa. G.B. (2003) Sustainability analysis of ecological and conventional systems in Bangladesh. *World Development* 31(10): 1721-1741.

Reijntjes, C., Bertus, H. & Water-Bayer, A. (1992) *Farming the future: An introduction to low external inputs and sustainable agriculture.* London: Macmillan.

Rigby, D., Woodhouse, P., Burton, M. & Young, T. (2001) Constructing a Farm-Level Indicator of Agricultural Sustainability.Sustainability. *Ecological Economics* 39: 463-478.

Rigby, D. & Caceres, D. (2001) Organic farming and the sustainability of agricultural system. *Agricultural Systems* 68: 21-40.

Risch, S. (1981) Insect herbivore abundance in tropical monocultures and polycultures: an experimental test of two hypotheses. *Ecology* 62: 1325-1340.

Rougoor, C.W., Ruud, G.T., Huirne, B.M. & Renkema, J.A. (1998) How to define and study farmers' management capacity: theory and use in agricultural economics. *Agricultural Economics* 18: 261-272.

Russillo, A. & Pintér, L. (2009) *Linking Farm-Level Measurement Systems to Environmental Sustainability Outcomes: Challenges and Ways Forward.* International Institute for Sustainable Development (IISD).

Ruthenberg, H. (1980) *Farming Systems in the Tropics,* 3rd edn. Oxford University Press

Sanchez, P.A. (1995) Science in agroforestry. *Agrofor. Syst.* 30: 5-55.

Sharma, S. (1996) *Applied multivariate techniques.* New York: John Wiley.

Shepherd, K.D. & Soule, M.J. (1998) Soil fertility management in west Kenya: dynamic simulation of productivity, profitability and sustainability at different resource endowment levels. *Agriculture, Ecosystems and Environment* 71: 131-145.

Sieber, N. (1999) Transporting the yield. Appropriate transport for agricultural production and marketing in Sub-Saharan Africa. *Transport Reviews* 19(3): 205-220.

Smaling, E. (1993) *An agro-ecological framework for integrating nutrient management, with special reference toKenya.* Agricultural University of Wageningen, the Netherlands, 250 p. (Ph.D. thesis).

Snapp, S.S., Blackie, M.J. & Donovan, C. (2003) Realigning research and extension to focus on farmers' constraintsand opportunities. *Food Policy* 28: 349-363.

Soil Survey Staff (1990) *Keys to soil taxonomy.* SMSS Technical Monograph No. 19, 4th ed. Virginia Polytechnic Institute and State University, USDA-SMSS.

Spenser, D.S.C. & Swift, M.J. (1991) Biodiversity and ecosystem function: Definition and measurement. In: Mulongoy, K., Gueye, M. & Spenser, D.S.C. (eds) *Biology nitrogen fixation and the sustainability of tropical agriculture.* Chichester: Wiley.

Stefanou, S.E. & Sexena, S. (1988) Education, experience and allocative efficiency: a dual approach. *American Journal of Agricultural Economics* 70: 338-345.

Tappan, G. & McGahuey, M. (2007) Tracking environment dynamics and agricultural intensification in southern Mali. *Agricultural Systems* 94(1): 38-51.

Tchabi, A., Coyne, D., Hountondji, F., Lawouin, L., Wiemken, A. & Oehl, F. (2008) Arbuscular mycorrhizal fungal communities in sub-Saharan Savannas of Benin, West Africa, as affected by agricultural land use intensity and ecological zone. *Mycorrhiza* 18: 181-195.

Tchienkoua, Z.W. (2003) Chemical and spectral characterization of soil phosphorus under three land uses from an Andic Palehumult in West Cameroon. *Agriculture, Ecosystems and Environment* 100: 193-200.

Trenbath, B.R. (1976) Plant interactions in mixed crop communities. In: Papendick, R.I., Sanchez, P.A., Triplett, G.B. (eds) *Multiple cropping. Madison: American Society of Agronomy*, pp. 129-169 (Special Publication, 27).

UN (2009) *World Population Prospects: The 2008 Revision.* New York: United Nations.

USDA (1999) *Sustainable Agriculture: Definitions and Terms.* Special Reference Briefs Series No. SRB, pp. 99-102.

Vandermeer, J. (1995) The ecological basis of alternative agriculture. *Annual Review of Ecology and Systematics* 26: 201-224.

Van der Pol, F. (1992) *Soil Mining. An Unseen Contributor to Farm Income in Southern Mali.* Bulletin of the Royal Tropical Institute No. 325, KIT Press, 47 p.

Vlaming, J.H., Van den Bosch, M.S., Van Wijk, A., De Jager, A.B. & Van Keulen, H. (2001) *Monitoring nutrientflows and economic performance in tropical forming systems (NUTMON).* Publishers: Alterra, Green World Research and Agricultural Economics Research Institute, LEI, The Netherlands.

Waithaka, M.M., Thornton, P.K., Herrero, M. & Shepherd, K.D. (2006) Bio-economic evaluation of farmers' perceptions of viable farms in western Kenya. *Agricultural Systems* 90: 243-271.

Wang, J., Wailes, E.J., Cramer, G.L. (1996). A shadow-price frontier measurement of profit efficiency in Chinese agriculture. *American Journal of Agricultural Economics* 78: 146-156.

Wakeyo, M.B. & Gardebroek, C. (2013) Does water harvesting induce fertilizer use among smallholders? Evidence from Ethiopia. *Agricultural systems* 114: 54-63.

Walker, T.S. & Jodha, N.S. (1986)How small farmers adapt to risk. In: Hazell, P., Pomareda, C. & Valdes, A. (eds) *Crop Insurance for Agricultural Development: Issues and Experience.* Baltimore, USA: John Hopkins University Press.

Webster, P. (1999) The challenge of sustainability at the farm level: presidential address. *Journal of Agricultural Economics* 50(3): 371-387.

Winrock International (1992) *Animal Agriculture in Sub-Saharan Africa: Executive Summary.* Winrock International Institute, Morrilton, USA.

Wolf, E.C. (1986) *Beyond the Green Revolution: New Approaches for Third World Agriculture.* Washington, DC: Worldwatch Institute, 46 p.

Wood, D. & Lenné, J.M. (1999) *Agrobiodiversity: Characterization, Utilization and Management.* Wallingford, UK: CABI, 490 pp.

Wopereis, M.C.S., Tamelokpo, A., Ezui, K., Gnakpenou, D., Fofana, B. & Breman, H. (2006) Mineral fertilizer management of maize on farmer fields differing in organic inputs in the West African savanna. *Field Crops Research* 96: 235-362.

Yang, D.T. (1997) Education in production: measuring labor quality and management. *American Journal of Agricultural Economics* 79(3): 764-772.

Yanni, G. (1997) Comparative regional geography in India and West Africa: Soils, Landforms and Economic Theory in agricultural development strategies. *The Geographical Journal* 163(1): 38-44.

4 Variation of Biodiversity in Sacred Groves and Fallows in the Western Highlands of Cameroon

C.M. Tankou[1]*, G.R. de Snoo[2], G. Persoon[3] and H.H. de Iongh[2]

Published in the African Journal of Ecology

1 Faculty of Agronomy and Agricultural Sciences, University of Dschang, P.O. Box 222 Dschang, Cameroon.
2 Institute of Environmental Sciences, Leiden University, P.O. Box 9518, 2300 RA Leiden, The Netherlands
3 Department of Anthropology, Leiden University, P.O. Box 9518, 2300 RA Leiden, The Netherlands.
* Corresponding author. cmtankou@yahoo.com; tel: (237) 77 66 03 04; fax: (237) 33 45 15 66

Abstract

This study was conducted in order to estimate species richness and diversity in different ecosystems and understand the floristic changes resulting from variation in abiotic factors in sacred groves. Vegetative assessment in quadrats revealed that the sacred groves were rich in plant genetic diversity composed of a total of 42, 65 and 82 ethno-botanical species of herbs, shrubs and trees respectively, of varied ecological and economic importance. The herbaceous α-diversity was significantly higher in the fallows than the sacred groves at low altitudes. The tree species richness was higher at low altitudes compared to high altitudes with tree ß-diversity increasing with altitude. Varying combinations of soil pH, total P, total K, CEC and slope percent were related to herbaceous species richness, herbaceous Shannon index and shrub species richness. Intensive land-use has completely changed the structure of the native vegetation and caused severe plant diversity losses, though some useful forage species have been introduced in the area. Habitat changes in the sacred groves may be driven by biophysical while a combination of human and biophysical factors could be considered in the case of rotational fallow vegetation.

Key words

Biodiversity, Western Highlands of Cameroon, sacred groves, rotational fallows, abiotic factors.

4.1   Introduction

Loss of biodiversity in the tropics is principally due to the destruction of habitat by anthropogenic activities (Wilson, 2000) especially the clearing of natural vegetation and conversion into agricultural cropland, harvesting non-timber forest products, selective extraction of plants and animals, biological invasion, and monoculture (Swamy *et al.*, 2000; Mishra *et al.* 2004; Sundarapandian *et al.*, 2005). Biodiversity strongly influences the provision of ecosystem services and therefore human wellbeing (Ma, 2005). Cameroon has a complex mosaic of diverse habitats, with moist, tropical forest dominating in the south and covering 54% of the country (UNEP-WCMC, 2003), montane forest and alpine savannah in the highlands, and sub-Sahelian savannah in the far north (Letouzey, 1968, 1985; White, 1983). These diverse habitats harbour more than 9,000 species of plants, 160 of which are endemic (UNEP-WCMC, 2003). In the Western Highlands of Cameroon (WHC) there is however a preponderance of patches of land still preserved as sacred groves because of strong religious beliefs held by the indigenous people. These sacred groves, rich in medicinal, rare, and endemic plants, are refugia for the relic flora of the region

(Whittaker, 1975; Jeeva *et al.*, 2007). The WHC is considered one of the major agricultural zones of Cameroon. Intensive land-use due to demographic pressure has led to major changes in the agro-ecosystem in the WHC including the reduction in biodiversity. Most of the research on biodiversity has been concentrated in the humid rainforest agro-ecological zone of Cameroon (Comiskey *et al.*, 2003) and virtually little or no attention paid to the species richness and diversity in the WHC. This paper aims at highlighting the biodiversity situation of this hitherto neglected zone with emphasis on the potential of the sacred groves and their differences with fallowed lands.

The primary determinants of change in species composition and community structure in undisturbed ecosystems are abiotic factors that vary with altitude (Whittaker, 1975). The establishment and management of a modified and simplified plant community, influences the composition and activities of the associated herbivore, predator, symbiont and decomposer sub-communities (Swift & Anderson, 1993). Timber exploitation and shifting cultivation have accounted for the destruction of biodiversity in the humid south of Cameroon (Zapfack *et al.*, 2002), while demographic pressure and human mobility have provoked fragmentation of large natural areas into small pockets in the WHC. This has resulted in increased intensity of land use over time and space and the permanent destruction of natural habitats that have greatly influenced plant species richness and diversity in the area. Fallow vegetation dominated by herbs and grasses that succeed several years of intensive food and cash crop production dominates most of the fragmented land cover in this zone which was previously occupied by a significant population of trees and shrubs (FAO-UNDP, 1979).

Species richness which represents the number of species in a given area (Reitalu *et al.*, 2009) is considered to be a prominent factor of productivity and stability (Cristofoli *et al.*, 2010; Gonga *et al.*, 2008). Species richness is predominantly controlled by local factors, and only secondarily by factors operating at the landscape level (Marini *et al.*, 2007). Quantifying the species richness for a site, landscape, or region is a practical way of describing plant community diversity and is a useful and most widely used measure for making comparisons among sites (Gotelli & Colwell, 2001). Species richness determines resource availability, growth conditions or the degree of impact from disturbance and resilience (Peet *et al.*, 1998; Fridley *et al.*, 2005). Many studies have observed that abiotic environmental factors, such as topographic (altitude, slope angle and aspect) and soil parameters can be important determinants of plant diversity (Bennie *et al.*, 2006; Marini *et al.*, 2007; Cristofoli *et al.*, 2010; Marini *et al.*, 2007). To understand the dynamics of population extinction and (re)colonization in fragmented landscapes, it is necessary to consider the cumulative effects of abiotic and biotic factors on the performance of adult plant species (Soomers, 2012).

Two types of agricultural biodiversity are identified by the Convention on Biological Diversity, a managed portion that is manipulated by people for their own needs and an unmanaged portion such as soil microbes, natural enemies, pollinators and their food plants that support production (Biodiversity International, 2007). Conventional practices have tended to promote a small number of species, and scientific research has typically been focused on these species (FAO, 2002), resulting in a decline in genetic diversity for agricultural crops. Many villages in the WHC have sacred groves which have great traditional values and contain contrasting biodiversity compared to the intensively used fragmented agricultural lands (Pélissier, 1980).The groves are repositories of biodiversity and harbour many threatened floral and faunal species and are the places where the village deity resides. Their ecological significance includes: conservation of biodiversity, recharge of aquifers, and soil conservation. Rotational fallows are the predominant land cover in the WHC especially during the dry season when the soil moisture content does not favour the cultivation of the common annual crops of the area. The ecosystems considered in this study are thus the sacred groves and the rotational fallows.

Theoretical explorations of the relationship between biotic and abiotic diversity are abundant (Huston, 1994; Rosenzweig, 1995). To gain an insight into the processes that may affect species distribution and diversity in the WHC, we analyzed the species diversity components with respect to abiotic environmental factors (altitude, slope angle, slope aspect, soil physical and chemical properties) and the basic ecosystems of the area. The main objective of the present study was to investigate the extent of tree, shrub and herbaceous plant species richness in the sacred groves, as dictated by topographical features and abiotic factors and quantify the impact of human disturbance through the evaluation of the herbaceous species in the fallowed lands with a view to generating baseline data of use to conservation. One of our hypotheses is that variation in altitude and abiotic factors contribute significantly to the explanation of species distribution. Another hypothesis is that the intensive land-use for agricultural purposes has had a significant effect on species composition of the different ecosystems.

The aims of this study were thus to:
- Identify and analyse botanical species composition and richness in the sacred groves
- Identify and analyse botanical species composition and richness in fallowed land
- Quantify the influence of abiotic factors and altitude on species diversity indices.

4.2 Materials and Methods

4.2.1 Study Area

The research was conducted in an area situated between the geographical co-ordinates 5° 27' - 5° 37.62' N and 09° 57.502' - 10° 09.544' E occupied by two sub-divisions in the Menoua Division (Fongo-Tongo and Nkong-ni) found in the WHC (Figure 4.1).

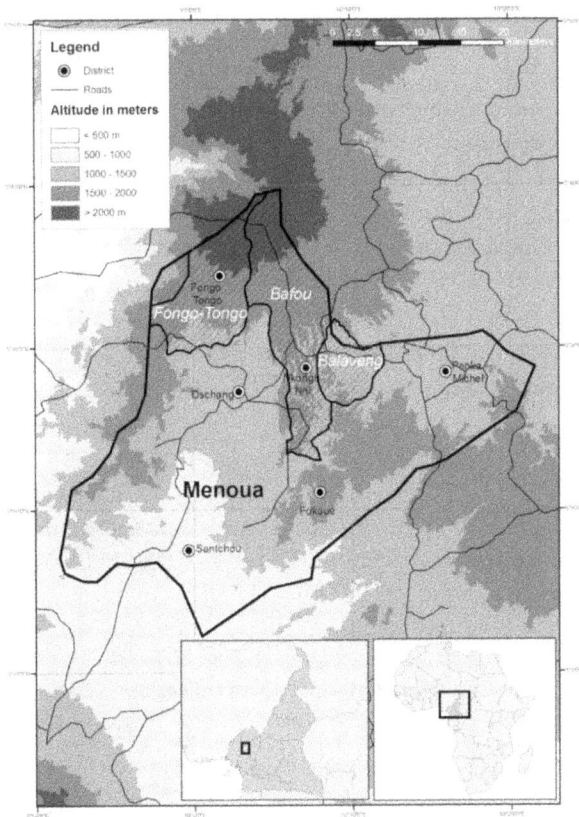

Figure 4.1
Geographical location of research site.

The characteristics of these two sub-divisions reflect the main features found in the WHC. The altitudinal range of the research sites are between 1400 m and 2500 m above sea level. The site has a tropical climate with a unimodal rainfall distribution. The growing season is between mid March and mid November and the dry season is between mid November and mid March. The annual rainfall is estimated to be between 1000 and 2000 mm (Kay *et al.*, 1985), with a mean annual temperature estimated at 20°C and the average annual sunshine estimated at 2000 hours. The soil is characterized by granite and gneisses in the southern lower altitudes and basaltic plateau at northern higher altitudes (Fotsing, 1992). The natural vegetation is dominated by grass withshrubs and trees. Many households have planted Eucalyptus trees (most often for border demarcation) which are commonly used as timber and wood for fuel. Remnants of what used to exist in this area can be found in the traditional sacred groves which are strictly out of bounds, especially to foreigners. These groves are believed to be the home of the ancestors. It is thanks to the fact that these groves are sacred and thus revered, that the miscellany of the original plant species of the area has been preserved.

Cool season vegetable crop production is dominant at higher altitudes and agricultural activities are also very common in inland valley swamps and on steep slopes due to land scarcity. Cropping systems vary according to altitude, with intercropping predominant at lower altitudes while at higher altitudes, sole cropping is more practiced and the crops (predominantly cool season vegetables) grown are more for the market. Fallowing is less common in the highly populated Bafou village while it lasts for between 2 and 5 years in some parts of Fongo-Tongo and Baleveng villages. Fallows are generally dominated by bush vegetation which is often exploited by cattle herders.

4.2.2   Data Collection

Data were collected at different altitudinal levels across the undisturbed (sacred groves) and disturbed vegetation (fallows) of the study area (Tables 4.1 and 4.2).

Table 4.1

Location of fallowed land used for biodiversity data collection.

Location	Geographic coordinates	Elevation (m)
Bafou (Femok)	5° 36′ N / 10° 2′ E	2355
Bafou (Femok)	5° 37′N/10° 2′ E	2430
Bafou (Femok)	5° 37.62′N/10° 1.88′E	2310
Bafou (Femok)	5° 33′N/10° 4′ E	1959
Bafou (Loung)	5° 33′N/10° 4′ E	1810
Bafou (Loung)	5° 33′N/10° 5′ E	1777
Baleveng	5° 30′ N/10° 8′ E	1523
Bafou (Nstingbeu)	5° 27′N/10° 7′ E	1516

Table 4.2

Location of sacred groves used for biodiversity data collection

Location	Geographic coordinates	Elevation (m)	Terrain slope (%) aspect
Baleveng (Fohnon)	5° 30.348′ N 10° 09.544′ E	1450	16.21 N
Bafou (Batsingla)	5° 27.332′ N 10° 06.6178′ E	1516	25.78 NW
Fongo-tongo (Mamiwatta)	5° 32.594′ N 09° 59.678′ E	1543	49.17 SE
Fongo-tongo (Apouh)	5° 27.332′ N 10° 06.617′ E	1597	33.70 NW
Fongo-tongo (Ndento)	5° 30.225′ N 09° 57.502′ E	1664	60.77 N
Fongo-tongo (Mepong)	5° 34.852′ N 10° 00.167′ E	1699	76.80 N
Fongo-tongo	5° 34.003′ N 09° 59.986′ E	1804	44.11 NE
Fongo-tongo	5° 34.185′ N 10° 00.281′ E	1809	43.37 N
Fongo-tongo (Apouh)	5° 34.133′ N 10° 00.166′ E	1812	47.51 NE
Fongo-tongo	5° 34.091′ N 10° 00.099′ E	1827	70.72 N
Fongo-tongo (Femok)	5° 35.153′ N 10° 00.957′ E	1981	42.91 N

At each level, we randomly sampled all major vegetation communities on fallows, using 1 m × 1 m quadrat. Trees and shrubs were not sampled in the fallows because they were virtually absent. In the sacred groves, data was collected randomly from 11 locations, each of which comprised a 0.0625 ha (25 m × 25 m quadrat) for trees with a diameter at breast height (DBH) greater than 10 cm, 0.01 ha (10 m × 10 m quadrat for shrubs or regenerated plants with DBH less than 10 cm and with a height of less than 3m. and 0.001ha (1 m × 1 m quadrat) for under storey herbaceous plants or ground-layer species. The research site was divided into two major groups with respect to altitude. Low altitude comprised all sites below 1800 m above sea level, while high altitude comprised all the sites above 1800 m above sea level. A total of six and five sacred groves were analysed for wild trees and shrubs in the low and high altitudes respectively while there were 12 and 11 samples collected in the low and high altitudes respectively for understorey herbaceous plants in the sacred groves. There were 18 and 28 samples collected in the fallow vegetation in the low and high altitudes respectively. Within and between altitudes, replications were thus carried out for the 1 m × 1 m data collection in the different systems, while within altitudes replications were used for the trees and shrubs data collection in the sacred groves.

Biological and morphological types were identified in the field using wild-flower, grass, and tree/shrub guides. Specimens of unidentified species were collected dried, and mounted in accordance with conventional herbarium practice for identification at the National Herbarium in Yaounde.

The altitude, percent slope and slope aspect of the plot were determined in the field.

Three replicate soil samples were taken per site at a depth of 0-30 cm in each of the sacred groves studied. The samples were homogenized by hand mixing and large live plant material (roots and shoots) and pebbles in each sample were separated by hand and discarded. The soil samples were air-dried and sieved for determination of soil factors through routine analyses in the Soil and Environmental laboratory in the University of Dschang, using methods described by Anderson and Ingram (1993).

The North-West and West Regions (Table 4.3) that make up this agro-ecological zone are the densest in the country as shown by the 2005 census results (Libite, 2010).

Table 4.3
Population data in 2005 and land area distribution in Cameroon.

Region	Population	Area (km²)	Number of Divisions	Population density (inhabitants/km²)
Adamawa	884289	63701	5	13.9
Centre	3098044	68953	10	44.9
East	771755	109002	4	7.1
Far North	3111793	34263	6	90.8
Littoral	2510363	20248	4	124
North	1687959	66090	4	25.5
North-West	1728953	17300	7	99.9
West	1720047	13892	8	123.8
South	634655	47191	4	13.4
South-West	1316079	25410	6	51.8

4.2.3 Analytical methods

Relative density and relative frequency were calculated according to Mori *et al.* (1983).

Relative frequency = 100 × (number of sample units containing a species/sample units for species of the sample)

Relative density = 100 × (number of individual of a species/number of individuals of a sample)

Species richness (S) was determined by the total number of species present in a sample. In order to evaluate the species diversity we used the Shannon-Weaver index, which represents α biodiversity showing local richness (Shannon & Weaver, 1949; Magurran, 1988; Frontier & Pichot-Viale, 1998; Vanpeene-Bruhier, 1998; Faurie, 2003; Gosselin & Laroussinie, 2004). The Shannon-Weiner index is:

$H = -\Sigma\,(p_i\,{}^*\ln p_i)$, where p_i, is the proportion of species *i* in the sample and ln the natural log function. The Shannon evenness index (J) was used to measure the equality of the abundance of each species. It was calculated as: $J = H / \ln(S)$ ß-diversity used to assess the environmental variability of trees in the sacred groves was measured as described by Balvanera *et al.* (2002) on the basis of the Jaccard similarity index,

$Cj = j/(a+ b- j)$

with *j* = number of species shared by two sites, *a* = number of species in site 1, and *b* = number of species in site 2. ß -diversity was defined here as:

$ß\,C_j = 1\text{-}\,C_j = (a+ b\text{-}2\,j)/(a+ b\text{-} j) = $ Number of species exclusive to *a and b/* Total number of species in both sites.

Clearly, ß-diversity is highest when the similarity is lowest. A correction factor was used to assess the role played by changes in species richness relative to changes in the amount of shared species in Jaccard's ß-diversity values:

$CF_c = C_{j[a = b]}/C_{j[a \neq b]} = (a+ b\text{-} j)/(2a\text{-}j)$

$C_{j[a = b]}$ or the Jaccard index when both sites had the same richness, and $C_{j[a \neq b]}$ or the Jaccard index when both sites had a different richness. In the calculations, *a* was used for the richness of the site with the largest number of species so that $CFc \leq 1$. A maximum value of $CFc = 1$ was obtained when both sites had the same species richness, and thus 100% of *Cj* was given by the number of shared species. *CFc*was lower as the differences between *a* and *b* were higher, so that the proportion of *Cj* given by the number of shared species was lower and that given by changes in species richness was higher.

ß -diversity was also measured using Whittaker's index,

$ß_w = γ/(α-1)$, where $γ = a + b - j$, for pair-wise comparisons only, and $γ$ = overall richness, for multiple site comparisons. Also, $α = (a + b)/2$ for pair-wise comparisons, and α= average richness of all sites, for multiple site comparisons.

Stepwise regression analysis was used to understand the effects of independent variables on dependent variables and the t-test was used to separate means between treatments.

4.3   Results

The total number of species, genera and families in the research area is summarized in Table 4.4.

Table 4.4

Species, genera and families recorded in the research area.

Treatment	Number of species	Number of families	Number of genera
Fallow herbs	76	24	56
Sacred grove herbs	42	26	38
Sacred grove shrubs	65	39	58
Sacred grove trees	82	38	63

The highest number of species and genera were found in the tree vegetation of the sacred grove and the highest number of families found in the shrub vegetation of the sacred grove. The fallow herbaceous species had the highest species/family and genera/family ratios of 3.17 and 2.33 respectively while the least with 1.62 and 1.46 respectively was found in the sacred grove herbaceous species. The species with the highest relative density (15.07) in the fallow ecosystem was *Sporobolus sp* and the species with the highest relative frequency (39.13) were *Ageratum conyzoides* and *Sporobolus sp*. The herbaceous species with the highest relative density (19.74) in the sacred grove was *Oplismenus hirtellus* and the one with the highest frequency (86.96) was *Phyllantus amarus*. The shrub species with the highest relative density (5.31) and relative frequency (54.55) in the sacred grove were *Dracaena diesteliana*, and *Bridelia*

---

micrantha. The tree species with the highest relative density (5.06) and frequency (45.97) was *Bridelia micrantha*.

4.3.1 Shared Species

There were six herbaceous species common in the fallow ecosystem and the sacred groves (Table 4.5) belonging to five families.

Table 4.5
Species of herbs in both fallows and sacred grove.

Species	Relative density in fallows	Relative density in sacred grove	Relative frequency in fallows	Relative frequency in sacred grove	Family
Aspilia Africana	5.01	0.52	13.04	8.70	Asteraceae
Chromolaena odora	0.03	0.52	2.18	8.70	Asteraceae
Commelina benghalensis	4.02	3.12	21.74	17.39	Commelinaceae
Desmodium hirtum	1.35	0.52	13.04	8.70	Fabaceae
Pteridium aquilinum	1.23	1.04	17.39	8.70	Hypolepidaceae
Sanicula elata	0.66	2.08	6.57	8.70	Apiaceae

There were 32 common species present both as shrubs and trees in the sacred grove belonging to 26 families (Table 4.6).

Table 4.6
Common species of trees and shrubs in the sacred groves.

Species	Relative density / frequency of trees	Relative density / frequency of shrubs	Family	Uses
Albizia adianthifolia (Schum.) W.F. Smith	3.93 / 35.75	3.54 / 36.36	Fabaceae	Erosion control and carving
Allophylus africanus	0.51 / 5.11	0.88 / 9.09	Sapindaceae	Toothache and diarrhoea
Allophylus bullatus	1.69 / 15.32	1.77 / 18.18	Sapindaceae	–
Bersama engleriana	1.12 / 10.21	1.77 / 18.18	Melianthaceae	Diabetes
Bridelia micrantha (Hochst.) Baill	5.06 / 45.97	5.31 / 54.55	Phyllanthaceae	Cough

Caloncoba glauca (P. Beauv.) Gilg.	1.12 / 10.21	0.88 / 9.09	Flacourtiaceae	Inflamation andskin diseases
Canthium mannii	1.69 / 15.32	1.77 / 18.18	Rubiaceae	Worms
Carapa grandiflora Sprague	1.12 / 10.21	1.77 / 18.18	Meliaceae	–
Cassipourea sp.	1.12 / 10.21	0.88 / 9.09	Rhizophoraceae	Timber
Cola acuminate	2.25 / 20.43	0.88 / 9.09	Sterculiaceae	Fatigue, Libido
Crassocephalum mannii	1.69 / 15.32	0.88 / 9.09	Asteraceae	Malaria
Croton oligandrum Pierre. Mss.	1.69 / 15.32	0.88 / 9.09	Euphorbiaceae	–
Cyathea manniana HK.	1.12 / 10.21	0.88 / 9.09	Cyatheaceae	–
Dracaena arborea	0.56 / 5.11	0.88 / 9.09	Asparagaceae	–
Ficus natalensis	1.12 / 10.21	0.88 / 9.09	Moraceae	Stomach disorder
Flacourtia flavescens Wild	1.68 / 15.32	1.77 / 18.18	Flacourtiaceae	Syphilis
Gambeya albida Aubrev et Pellegr.	0.56 / 5.11	0.88 / 9.09	Sapotaceae	–
Macaranga occidentalis	2.25 / 20.43	0.88 / 9.09	Euphorbiaceae	–
Maesa lanceolata Forsk	2.81 / 25.54	1.77 / 18.18	Myrsinaceae	Tape worm
Nuxia congesta R. Br. Ex Presen.	2.25 / 20.43	2.65 / 27.27	Stilbaceae	–
Olea capensis Linn	0.51 / 5.11	0.88 / 9.09	Oleaceae	–
Persea americana Mill	2.81 / 25.54	0.88 / 9.09	Lauraceae	Hypertension and diabetes
Polyscias fulva	2.81 / 25.54	1.77 / 18.18	Araliaceae	Headache
Protea argyrophaea Hutch	2.25 / 20.43	3.54 / 36.36	Proteaeceae	–
Pseudospondias microcarpa (A. Rich.) Engl.	2.25 / 20.43	2.65 / 27.27	Anacardiaceae	Syphilis
Psorospemum kunthianum	0.56 / 5.11	0.88 / 9.09	Clusiaceae	–
Psychotria peduncularis	1.12 / 10.21	4.42 / 45.45	Rubiaceae	–
Rauwolfia vomitoria	2.25 / 20.43	2.65 / 27.27	Apocynaceae	Hypertension
Salacia mayumbensis N. Hallis	1.12 / 10.21	1.77 / 18.18	Celastraceae	–
Trema guineensis (Schum & Thonn.) Ficalho	1.69 / 15.32	1.77 / 18.18	Cannabaceae	Hypertension and diabetes
Tricalysia biagrana Heirn	2.25 / 20.43	5.31 / 54.54	Rubiaceae	–
Vitex doniana	0.56 / 5.11	0.88/ 9.09	Verbenaceae	Epilepsy

4.3.2 Unique Species

There were 35 herbaceous species belonging to 23 families found in the sacred groves and absent in the fallow vegetation (Table 4.7) while 70 herbaceous species belonging to 23 families found in the fallow vegetation were absent in the ground layer of the sacred groves (Table 4.8).

Table 4.7

Species of herbs in the ground-level sacred grove absent in the fallow vegetation.

Species	Family	Relative density	Relative frequency
Adenia cissampeloides	Passifloraceae	1.56	8.70
Aframomum aulacocarpus	Zingeberaceae	0.51	8.70
Aframomum danielli (Hook. f.) K. Schum.	Zingeberaceae	2.34	21.74
Arthropteris palisoti	Oleandraceae	1.82	21.74
Astystasia sp.	Acanthaceae	2.08	8.70
Asystasia gangetica	Acanthaceae	17.14	17.39
Begonia adpressia	Begoniaceae	7.79	4.35
Brillantaisia nitens Lindau	Acanthaceae	0.52	8.70
Clerodendron sp.	Lamiaceae	0.52	8.70
Crassocephalum biafrae	Asteraceae	0.26	4.35
Craterosiphon scandens	Thymelaeaceae	0.52	8.70
Cyathea manniana HK.	Cyatheaceae	0.52	8.70
Desmodium velutinum	Fabaceae	0.52	8.70
Dioda scandens SW	Rubiaceae	2.86	8.70
Eremomastax speciosa (Hochst.) cufod	Acanthaceae	1.56	8.70
Gouania longipetala Hemsl	Rhamnaceae	1.04	13.04
Justicia depauperata	Acanthaceae	2.86	26.09
Laggera pterodonta (D.C) Sch.	Asteraceae	0.26	4.35
Laportea ovalifolia	Urticaceae	3.38	13.04
Melinus minutiflora	Poaceae	5.19	8.70
Microglossa angolensis Oliv.	Asteraceae	0.52	8.70
Momordica foetida Schumach.	Curcubitaceae	1.04	13.04
Oplismenus hirtellus	Graminae	19.74	52.17
Palisota barteri	Commelinaceae	0.78	4.35
Paullinia pinnata	Sapindaceae	0.78	13.04

Phyllantus amarus	Euphorbiaceae	0.52	86.96
Piper capensis	Piperaceae	3.38	13.04
Rubus idaeus	Rosaceae	0.52	8.70
Setaria faberi Herrm	Poaceae	0.26	4.35
Smilax kraussiana	Smilacaceae	0.52	8.70
Spermacoce princeae k. Schum	Rubiaceae	7.79	8.70
Spermacoce saticola k. Schum.	Rubiaceae	0.26	4.35
Tectaria angelicifolia	Tectariaceae	0.52	8.70
Tragia benthami Bak	Euphorbiaceae	0.26	4.34
Urera gabonensis Pierre ms	Urticaceae	0.26	4.34

Table 4.8
Fallow species not found in the sacred groves.

Species	Family	Relative density	Relative frequency
Acanthus montanus	Acanthaceae	0.24	4.34
Acanthus sp.	Acanthaceae	0.06	2.17
Achyranthes aspera	Amaranthaceae	0.18	4.35
Achyranthes sp.	Amaranthaceae	0.45	2.18
Ageratum conyzoides	Asteraceae	8.58	39.13
Ageratum houtonium	Asteraceae	0.63	4.35
Ageratum sp.	Asteraceae	0.42	4.35
Amaranthus spinosus	Amaranthaceae	0.12	6.52
Aneilema commelina	Commelinaceae	0.51	4.31
Aneilema sp.	Commelinaceae	0.42	4.34
Bidens pilosa	Asteraceae	4.29	30.43
Bidens spinosa	Asteraceae	0.48	6.52
Caucalis platycarpos	Apiaceae	0.06	4.34
Colocassia esculenta	Araceae	1.17	2.17
Comelina sp.	Commelinaceae	0.12	2.17
Crepis sp.	Asteraceaae	0.21	6.52
Crotalaria irsuta	Fabaceae	0.09	2.17
Crotalaria sp.	Fabaceae	4.23	8.7
Cynodon dactylon	Poaceae	5.37	21.74
Cyperus distans	Cyperaceae	0.06	2.17
Cyperus sp.	Cyperaceae	1.38	10.87

Desmodium trifolia	Fabaceae	0.09	2.17
Dichrocephala integrifolia	Asteraceae	0.45	2.17
Diodia sanders	Rubiaceae	0.54	2.17
Dioscorea sp.	Dioscoreaceae	0.03	2.17
Echinops giganteus	Asteraceae	0.06	4.35
Emilia coccinea	Asteraceae	0.12	2.17
Erigeron floribundus	Asteraceae	3.09	26.09
Erigeron sp.	Asteraceae	0.54	6.52
Eriosema linifolium	Fabaceae	0.06	2.17
Euphorbia sp.	Euphorbiaceae	0.15	2.17
Galinsoga sp.	Asteraceae	0.06	2.17
Helichrysum rutidolepis	Asteraceae	3.00	13.04
Helichrysum cymosun	Asteraceae	0.27	6.52
Hydrocotyle manii	Araliaceae	0.06	2.17
Hypericum lanceolatum	Hypericaceae	0.30	6.52
Imperata cylindrical	Poaceae	3.48	13.04
Imperata sp.	Poaceae	0.81	6.52
Kosteletzkya sp.	Malvaceae	0.12	2.17
Ludwigia sp.	Onagraceae	1.11	4.35
Mariscus alternifolius Vahl.	Cyperaceae	0.03	2.17
Mimosa sp.	Fabaceae	0.69	4.35
Mitracorpus villosus	Rubiaceae	0.15	2.17
Musa sp. (AB)	Musaceae	0.03	2.17
Musa sp. (AA)	Musaceae	0.18	4.35
Oxalis corniculata	Oxalidaceae	2.91	4.35
Oxalis sp.	Oxalidaceae	3.45	26.09
Paspalum sp.	Poaceae	0.03	2.17
Pennisetum clandestinum	Poaceae	6.78	28.26
Pennisetum purpureum	Poaceae	1.68	15.22
Persea Americana	Lauraceae	0.12	6.52
Polystichum sp.	Dryopteridacea	1.32	10.87
Satureja robusta	Labiatae	0.06	4.35
Sida acuta	Malvaceae	1.80	17.39
Sida rhombifolia	Malvaceae	0.69	4.35

Sida sp.	Malvaceae	0.06	2.17
Siegesbeckia sp.	Asteraceae	1.26	0.04
Solanum scabrum	Solanaceae	1.26	4.35
Splilanthes filicaulis	Asteraceae	1.23	6.52
Sporobolus sp.	Poaceae	15.07	39.13
Stellaria sp.	Caryophyllaceae	2.82	8.70
Tithonia diversifolia	Asteraceae	0.09	2.17
Tradescantia sp.	Commelinaceae	0.87	4.35
Triplasis sp.	Poaceae	0.03	2.17
Urena lobata	Malvaceae	0.75	8.70
Vernonia amygdalina	Asteraceae	0.15	6.52
Vernonia calvoana	Asteraceae	1.32	13.04
Waltheria indica	Malvaceae	0.15	2.17
Xanthosoma sagitifolium	Araceae	0.45	2.17
Xanthosoma sp.	Araceae	0.03	2.17

In addition, there were 32 unique shrub species belonging to 19 families (Table 4.9) and 49 unique tree species belonging to 28 families in the sacred groves (Table 4.10). The species found in the sacred groves possessed varied potentials including land conservation, timber and especially medicinal properties (Tables 4.6, 4.9 and 4.10).

Table 4.9
Shrub species not found as trees in scared groves

Species	Family	Relative density	Relative frequency	Uses
Acalypha manniana	Euphorbiaceae	0.88	9.09	Skin diseases
Acioa johurtonei Hoyle	Chrysobalanaceae	0.88	9.09	–
Albizia sp.	Fabaceae	0.88	9.09	–
Alchornea laxiflora	Euphorbiaceae	0.88	9.09	–
Antidesma sp.	Phyllanthaceae	1.77	18.18	–
Brucea antidysenterica	Simarounbaceae	1.77	18.18	Dysentery
Cassine aethiopica	Celastraceae	0.88	9.09	
Chrysobalanus icaco	Chrysobalanaceae	0.88	9.09	–
Clausena anisata	Rutaceae	0.88	9.09	Cough, cold

Clutia kamerunica	Euphorbiaceae	0.88	9.09	–
Dalbergia oligophylla Hutch & Dalziel	Fabaceae	1.77	18.18	–
Diphasia angolensis	Rutaceae	0.88	9.09	–
Dracaena diesteliana	Agavaceae	5.31	54.55	–
Elaeis sp.	Arecaceae	0.88	9.09	–
Ficus mucoso	Moraceae	0.88	9.09	Insomnia
Gnidia glauca	Thymelaeaceae	0.88	9.09	–
Leea guineensis	Leeaceae	1.77	18.18	Toothache
Neoboutonia glabrescens	Euphorbiaceae	0.88	9.09	Worms
Peddiea fischeri	Thymelaeaceae	0.88	9.09	–
Piper capensis	Piperaceae	1.77	18.18	
Pittosporum mannii Hook. f.	Pittosporaceae	1.77	18.18	Intestinal diseases
Psorospermum cf. aurantiacum Engl.	Clusiaceae	1.77	18.18	–
Psorospermum senegalensis Spach.	Clusiaceae	0.88	9.09	–
Psychotria camtopus	Rubiaceae	0.88	9.09	–
Psychotria cf djumaensis	Rubiaceae	0.88	9.09	–
Raphia sp.	Arecaceae	1.77	18.18	–
Stombosiopsis tetrandra	Olacaceae	0.88	9.09	–
Tarenna baconioides	Rubiaceae	0.88	9.09	–
Tephrosia vogelii Hook. f.	Fabaceae	0.88	9.09	Diarrhea
Tilia sp.	Malvaceae	0.88	9.09	–
Vernonia amygdalina	Asteraceae	0.88	9.09	Malaria
Xymalos monospora	Monimiaceae	0.88	9.09	–

Table 4.10

Tree species not found as shrubs in scared groves

Species	Family	Relative density	Relative frequency	Uses
Acioa sp.	Chrysobalanaceae	0.56	5.11	–
Agauria salicifolia	Ericaceae	0.56	5.11	–
Alangium chinense	Alangiaceae	2.81	25.54	Skin diseases
Albizia gummifera	Fabaceae	0.56	5.11	Malaria
Albizia zygia	Fabaceae	0.56	5.11	Depression
Allophylus hamatus	Sapindaceae	1.12	10.21	
Bersama abysinica	Melianthaceae	1.12	10.21	
Bridelia sp.	Phyllanthaceae	0.56	5.11	–
Campylostemon sp.	Celastraceae	0.56	5.11	–
Canarium schweinfurthii	Sapotaceae	0.56	5.11	Gonorrhoea
Canthium subcordatum	Rubiaceae	1.69	15.32	–
Cassine sp.	Celastraceae	1.12	10.21	–
Cordia auranthiata	Boraginaceae	1.12	10.21	
Cordia platythirsa Bak.	Boraginaceae	0.56	5.11	
Croton macrostachys	Euphorbiaceae	0.56	5.11	
Cussonia sp.	Araliaceae	0.56	5.11	–
Entandrophragma angolense	Meliaceae	0.56	5.11	Malaria
Fagara tessmannii	Rutaceae	0.56	5.11	–
Fagara zanthozyloides	Rutaceae	0.56	5.11	–
Ficus exasperate	Moraceae	1.12	10.21	Scabies
Ficus mucuso	Moraceae	1.69	15.32	Insomnia
Ficus sp.	Moraceae	0.56	5.11	–
Ficus thonningii	Moraceae	0.56	5.11	Wounds
Ficus tricopoda	Moraceae	2.25	20.43	–
Flacourtia sp.	Flacourtiaceae	0.56	5.11	–
Garcinia smeathmannii	Clusiaceae	0.56	5.11	
Holarrhena floribunda	Apocynaceae	0.56	5.11	Jaundice
Kigelia Africana	Bignomiaceae	1.12	10.21	Hemorrhage
Landolphia sp.	Apocynaceae	0.56	5.11	–
Magnistipula conrauana	Chrysobalanaceae	0.56	5.11	–

Mangifera indica	Anacardaceae	0.56	5.11	High blood pressure
Markhamia tomentosa	Bignoniaceae	1.12	10.21	Oedema
Octolepis casearia	Thymeleaceae	0.56	5.11	–
Olea hochstetteri	Oleaceae	0.56	5.11	Diabetes
Phoenix reclinata Jacq	Arecaceae	1.69	15.32	–
Pittosporum mannii Hook. f.	Pittosporaceae	2.25	20.43	Intestinal diseases
Prunus Africana	Rosaceae	0.56	5.11	Fungal and bacterial infections, madness
Psidium guajava	Myrtaceae	0.56	5.11	Diarrhea
Rothmannia urcelliformis	Rubiaceae	1.23	10.21	–
Salacia staudtiana	Celastraceae	0.56	5.11	–
Sapium ellipticum (Hochst.) Pax	Euphorbiaceae	2.25	20.43	Purgative and eczema
Schefflera barteri (Seem.) Harms	Araliaceae	0.56	5.11	–
Syzygium aromaticum	Myrtaceae	0.56	5.11	Mouth and teeth pain
Tricalysia atherura	Rubiaceae	0.56	5.11	Jaundice
Trichilia sp.	Meliaceae	1.69	15.32	–
Turraea vogelii	Meliaceae	0.56	5.11	Cough
Vepris louisii	Rutaceae	0.56	5.11	Malaria
Vernonia sp.	Asteraceae	0.56	5.11	–
Zanha golungensis	Sapindaceae	0.56	5.11	Wounds

4.3.3 Species richness and diversity in different systems

The herbaceous species richness and diversity recorded in the fallow vegetation and the sacred groves were compared. A significantly ($p < 0.05$) higher value of species richness was observed in the fallow system at the low altitude while there was no significant difference in the Shannon indices. At the high altitude, no significant differences were observed for the biodiversity indicators. The Shannon evenness indices for both the low and high altitudes were not significantly different (Table 4.11).

Table 4.11

Species richness and diversity of herbs in different ecosystems.

Variable	Means in the low altitude		Df	t-value[a]
	Fallow land n = 18	Sacred grove n = 12		
Species richness	6.00	3.75	28	2.66*
Shannon index	1.26	0.99	28	1.40 ns
Shannon evenness	0.74	0.65	28	0.84 ns
	Means in the high altitude			
	N = 28	N =11		
Species richness	5.96	5.36	37	0.99 ns
Shannon index	1.32	1.17	37	1.36 ns
Shannon evenness	0.76	0.71	37	1.03 ns

[a] * indicates significant at 5% probability level, ns indicates non significant (p > 0.05).

4.3.4 Variation within ecosystems

The effect of altitude on the different ecosystems was noticed only with tree vegetation of the sacred groves where both species richness and Shannon index were significantly higher at the lower than at the higher altitudes while all the variables were not significantly different for the shrub and herbaceous species (Table 4.12)

Table 4.12

Species richness and diversity within systems

Variable	Means				df	t-value[a]
	Low altitude	N	High altitude	n		
Species richness of trees in sacred grove	19.67	6	12	5	9	4.44**
Shannon index of trees in sacred grove	2.61	6	1.98	5	9	2.64*
Shannon evenness of trees in sacred grove	0.88	6	0.80	5	9	1.28ns
Species richness of shrubs	11.00	6	9.40	5	9	0.71 ns
Shannon index of shrubs in sacred grove	1.64	6	1.64	5	9	0.02 ns
Species richness of herbs in sacred grove	3.75	12	5.36	11	21	1.92 ns

Shannon index of herbs in sacred grove	0.99	12	1.17	11	21	0.78 ns
Shannon evenness of herbs in sacred grove	0.65	12	0.71	11	21	0.45ns
Species richness of herbs in fallows	6.00	18	5.96	28	42	0.06 ns
Shannon index of herbs in fallows	1.26	18	1.32	28	42	0.55 ns
Shannon evenness of herbs in fallow	0.75	18	0.76	28	42	0.34ns

[a] ** and * indicates significant at 1% and 5% probability levels respectively, ns indicates non significant ($p > 0.05$).

4.3.5 Variation of soil chemical and physical properties with altitude

Significantly higher values of organic matter and CEC were recorded at the higher altitude while significantly higher values of total K were found at the lower altitude (Table 4.13).

Table 4.13
Soil chemical and physical properties

Variable	Means		Df	t-value[a]
	Lowlands (n = 12)	Highlands (n = 11)		
pH	4.72	4.61	19	1.90 ns
Organic matter (%)	11.75	14.60	19	3.54**
CEC (cmol+/kg)	24.13	30.68	19	4.28**
Total N (%)	2.57	3.20	19	1.66 ns
Total P (ppm)	4.87	5.58	19	1.37 ns
Total K (‰)	0.08	0.06	19	3.35**
Sand (%)	17.40	10.36	19	1.34 ns
Silt (%)	68.00	77.82	19	1.24 ns
Clay (%)	15.00	12.00	19	0.65 ns

[a] ** indicates significant at 1% probability levels, ns indicates non significant ($p > 0.05$).

4.3.6   Influence of environmental factors on biodiversity indices.

The combination of abiotic factors was tested for all life forms. Statistically significant models were only detected for shrubs and herbaceous sacred grove species. Decreasing soil pH and total P increased the Shannon index of herbs in the sacred groves while increasing CEC and slope increased species richness of herbs in the sacred groves and decreasing total P, decreasing pH and increasing K increased species richness of shrubs in the sacred groves (Table 4.14).

Table 4.14
Significant stepwise regression results between biodiversity indices and environmental factors

Dependent variable	Model	Coefficients	F	R square
Shannon index of herbs in sacred grove	Constant	15.28	10.15**	0.53*
	pH	-2.82		
	Total P	-1.90		
Species richness of herbs in sacred grove	Constant	-3.64	14.55**	0.62*
	CEC	0.21		
	Slope	0.06		
Species richness of shrubs in sacred grove	Constant	129.77	13.24**	0.87*
	Total P	-3.26		
	pH	-23.35		
	Total K	938.14		

** and * indicates significant at 1% and 5% probability levels respectively.

The slope aspect had no significant effect ($p > 0.05$) on all the variables examined.

4.3.7   Assessment of altitudinal variability of trees

In order to gain more insight into the altitudinal variation shown in table 4.5, ß-diversity was examined for the trees in the sacred grove. Average $ßC_j$ and $ß_w$ were 0.94 and 2.02 respectively. $ßC_j$ and $ß_w$ showed similar patterns and were correlated ($r = 0.94$, $p < 0.001$). Generally the largest ß-diversity values ($ßC_j$ and $ß_w$) indicated comparisons between the poorest sacred grove and the richest one which were related to lowest values of $CFCj$. The poorest sacred grove was that found at altitude 1809 m asl with 10 species recorded while the richest was found at altitude 1664 with 25 species recorded $ßC_j = 0.97$, $ß_w = 0.69$ and $CFCj = 2.06$). Generally from the table it was observed that the values of ß-diversity increased gradually with increasing altitude (Table 4.15).

Table 4.15

ß-diversity measures among all pairs of 0.0625-ha quadrats of trees in the sacred groves using Jaccard similarity index (ßCj, in bold), Jaccard's correction factor (CFCj) (separated by /) and Whittaker's ß -diversity index (ßw, in parentheses)

Altitude (m)	1450	1516	1543	1597	1664	1699	1804	1809	1812	1827	1981
1450		.82/.90 (1.81)	.88/.97 (1.89)	.81/.94 (1.78)	.74/.83 (1.66)	.89/.92 (1.89)	.89/.82 (1.93)	.92/.77 (2)	.97/.86 (2.07)	.83/.75 (1.85)	.94/.91 (2)
1516			.94/.89 (2)	.94/.87 (2)	.82/.77 (1.79)	1/.86 (2.12)	.96/.9 (2.08)	1/.83 (2.17)	1/.93 (2.15)	.91/.82 (2)	1/1 (2.14)
1543				.89/.97 (1.89)	.87/.87 (1.86)	.95/.95 (2)	.97/.81 (2.07)	1/.76 (2.15)	1/.84 (2.13)	1/.76 (2.15)	1/.89 (2.13)
1597					.90/.89 (1.91)	.98/.98 (2.05)	.97/.79 (2.07)	.97/.74 (2.07)	1/.83 (2.13)	1/.75 (2.14)	1/.88 (2.12)
1664						.93/.91 (1.95)	.91/.72 (1.94)	.97/.69 (2.06)	1/.76 (2.11)	.94/.69 (2)	.97/.8 (2.05)
1699							1/.79 (2.13)	1/.74 (2.14)	.97/.8 (2.06)	1/.74 (2.14)	1/.86 (2.12)
1804								.84/.91 (1.9)	.86/.96 (1.91)	.9/.91 (2)	.96/.90 (2.08)
1809									.85/.87 (1.9)	.95/1 (2.11)	1/.83 (2.17)
1812										.9/.88 (2)	.96/.93 (2.08)
1827											.96/.83 (2.09)
No species	18	15	19	20	25	21	12	10	13	10	15

4.4 Discussion

The results of this research show that the WHC harbours significant numbers of species of different life forms. Fewer families nearly all of which were herbaceous species were found in the disturbed fallow ecosystem compared to the un-disturbed sacred grove ecosystem with trees and shrubs. The results also show that the WHC harbours a significant number of species with diverse importance in the sacred groves that include pharmaceutical substances, timber, firewood and raw materials for artistic works. This ties in with the findings that tropical vegetation are important as timber sources (Finegan, 1992), providers of environmental services such as protection from erosion and atmospheric carbon fixation, templates for forest rehabilitation (Lugo, 1992), refugia of biodiversity in fragmented landscapes (Lamb *et al.*, 1997), and as local providers of medicinal and useful plants (Lamb *et al.*, 1997).

4.4.1 Effect of altitude on biodiversity

At the lower altitudes, herbaceous species diversity was significantly higher in the fallow ecosystem than in the sacred grove indicating that the species pool in the sacred grove was limited by some natural factors. Differences in herbaceous species-type between the fallows and the sacred grove could be linked to the shading effect in the sacred grove which was not favourable to heliophilic species such as those of the Poaceae family. This explained the near nonexistence of C_4 herbaceous species in the sacred grove. The high altitude areas are very suitable for cool season crops which are the principal sources of revenue for rural farmers in the zone. Due to intense rural-rural mobility, the natural vegetation had, to a large extent been replaced by agriculture. The intensive land-use in the WHC has consequences for soil erosion and soil quality. In natural ecosystems, the vegetative cover of a forest or grassland has been shown to prevent soil erosion, replenish ground water and control flooding by enhancing infiltration and reducing water runoff (Perry, 1994). These processes depend upon the maintenance of biological diversity (Altieri, 1994). When the ecosystem services of biodiversity are lost due to biological degradation, the economic and environmental costs can be quite significant. Often the costs involve a reduction in the quality of life due to decreased soil, water, and food quality orchestrated by the negative effects of chemical inputs. The net result is an agro-ecosystem which is a man-made ecosystem that requires constant human intervention, whereas in natural ecosystems the internal regulation is a product of plant biodiversity through flows of energy and nutrients, and this form of control is progressively lost under agricultural intensification (Swift & Anderson, 1993). Despite the fact that intensive land-use had eroded some natural species, other important species have been introduced such as *Pennisetum clandestinum* which is an important pasture species for dairy cattle and is also widely used when laying a lawn. However it should be noted that the sacred grooves harbour a high biodiversity of trees, shrubs and herbaceous plants compared to the fallowed lands of the study area where trees and shrubs have been greatly eliminated.

Amongst all the life forms studied, significant differences in biodiversity with altitude were recorded only for tree vegetation. The trend of tree biodiversity variation was made clearer with the ß-diversity analyses. Givnish (1999) observed that describing and exploring the determinants of ß-diversity was particularly critical for tropical vegetation. The number of tree species decreased with altitude. Similar findings were shown by Ohlemüller and Wilson (2000). Variation of species richness with altitude had been greatly attributed to lower temperature at higher altitudes that were not favourable to many species. Rawat (2011) found that the number of species reduced at a rate of 3–4% per 100 m elevation in the high altitudes of Garhwal Himalaya due to increasing stress (lowering of temperature and pressure, thinning of air, etc.) amongst other factors. However, Khem & Ole (2003) stated that climate-elevational gradient of plant species richness was group-specific and varied according to life forms.

4.4.2 Variation and influence of abiotic factors

The soil organic matter content showed significant increase with altitude. The organic matter content could be linked to the varying rate of decomposition imposed by temperature where decreasing temperature with altitude decreased the decomposition rate and hence resulted in more accumulation at higher altitudes. Similar results were shown by Trumbore et al. (1996) who observed that light-fraction organic carbon increased exponentially with increasing temperature. CEC is the maximum quantity of total cations of any class, that a soil is capable of holding at a given pH value, for exchange with the soil solution. It was thus used as a measure of fertility. Higher altitudes were thus more fertile than the lower altitudes due to their high organic matter content. The potassium content was lower at the high altitudes than at the low altitudes. The higher the amount of exchangeable base cations, the more acidity could be neutralized. This explained why there was a decrease in soil acidity conditions at lower altitudes compared to higher altitudes. Though individual abiotic factors did not show any significant relationships with biodiversity indices measured, Hooper et al., 2005 revealed that ecosystem properties were apparently more influenced by abiotic conditions than by species richness. Spatial heterogeneity in the physical environment (e.g. substrate, nutrients, soil moisture and structure) has been positively and linearly correlated with diversity at a number of spatial scales (Harman, 1972; Schlosser, 1982; Tonn & Magnuson, 1982; Crozier & Boerner, 1984; Chambers & Prepas, 1990; Kaczor & Hartnett, 1990; Pringle, 1990; Scarsbrook & Townsend, 1993).

Combined abiotic factors explained the herbaceous and shrub biodiversity of the WHC. Soil pH and phosphorus decreased the Shannon index of herbaceous sacred grove species while CEC and slope increased the herbaceous species richness. Decreasing pH and phosphorus with increasing potassium increased shrub species richness. This is vital information as large-scale national biodiversity hotspots identified for plant species using abiotic methods, can become the nuclei for further detailed conservation planning (Venevsky & Venevskaia, 2005). John et al. (2007) analysed the variation in soil nutrients in three neo-tropical forest plots and concluded that the distribution of 40% of the species showed affinities with soil nutrients.

The sacred grove is a great treasure to the inhabitants of the WHC. They are the living granary of medicinal plants in the region. Forty-two species of medicinal plants were identified in the study area and similar results have been reported by Zapfack *et al.* (2002) in the forest zone of Cameroon.

4.5   Conclusion

The present study of plant communities with respect to ecosystems and abiotic factors at the local scale leads to valuable insights in resource partitioning and niche differentiation of the area studied. Land use is among the most important determinants of biodiversity as shown by the data collected in the WHC. Elevation influences the air temperature, its daily fluctuations, and the humidity levels, and all of these in turn influence the growth and development of plants. Altitudinal gradient influenced the partitioning of the tree species in the study area. Analysis of the abiotic factors showed that both fertility and soil pH increased with altitude and accounted for the richness of herbaceous species richness. Habitat changes in the sacred groves may be driven by biophysical factorss while in the rotational fallow vegetation it could be linked to a combination of human and biophysical factors.

Increase in human population numbers, causing loss of natural habitat and promoting the invasion of less desirable species such *Sporobolus sp* which are known to decrease pasture productivity, and decreasing the population of highly valued medicinal plants such as *Prunus africana* could be considered as the greatest single threat to species diversity in the region. Mobility within the rural communities in search of additional farming space, higher value arable land, and land for construction of houses in addition to the destruction of habitat through the creation of new roads, significantly aggravates plant diversity losses.

While intensive land use has deprived the area of much of the profit that can be derived from shrub and tree biodiversity depicted by the potential of the sacred groves, some valuable forage species have colonized most of the uncultivated portions. Given the importance of the conservation of biodiversity and the ecosystem, attempts should be made to maintain the sanctity of the sacred groves where there exist a high biodiversity of trees, shrubs and herbaceous plants that can hardly be found in the fallow vegetation and cultivated lands of the study area.

Acknowledgements

This research was made possible by funding from Volkswagen foundation. The authors wish to thank Maarten van't Zelfde of the Institute for Environmental Sciences (CML) Leiden, for producing the map of the research site and Tacham Walter of the University of Bamenda Cameroon, for help with the field work.

References

Altieri, M.A. (1994) Biodiversity and Pest Management in Agro ecosystems. New York: Haworth Press, 185 p.

Anderson, J.M. & Ingram, J.S. (1993) *Tropical Soil Biology and Fertility: Handbook and Methods*, 2nd Edition. CAB International, Wallingford, UK, 221 p.

Balvanera, P., Lott, E., Segura, G., Siebe, C. & Islas, A. (2002) Betadiversity patterns and correlates in a tropical dry forest of Mexico. *J. Veg. Sci.* 13: 145-158.

Bennie, J., Hill, M., Baxter, R. & Huntley, B. (2006) Influence of slope and aspect on long-term vegetation change in British chalk grasslands. *Journal of Ecology* 94: 355-368.

Biodiversity International (2007) *Sub-Saharan Africa*. Rome: Biodiversity Int.

Chambers, P.A, & Prepas, E.E. (1990) Competition and coexistence in submerged aquatic plant communities: the effects of species interactions versus abiotic factors. *Freshwater Biology* 23: 541-550.

Comiskey, J.A., Sunderland, T.C.H. & Sunderland-Groves, J.L. (2003) *Takamanda: the Biodiversity of an African Rainforest*. SI/MAB Series #8. Washington, DC: Smithsonian Institution.

Cristofoli, S., Monty, A. & Mahy, G. (2010) Historical landscape structure affects plant species richness in wet heath lands with complex landscape dynamics. *Landscape and Urban Planning* 98: 92-98.

Crozier, C.R. & Boerner, R.E.J. (1984) Correlations of ecol 79 114 understory herb distribution patterns with microhabitats under different tree species in a mixed mesophytic forest. *Oecologia* 62: 337-343.

FAO (2002) *The state of agricultural biodiversity in the livestock sector: Threats to livestock genetic diversity*. Rome: FAO.

FAO-UNDP (1979) *Soil Science Project. Soil survey and Land Evaluation for the Selection of land for seed multiplication farm*. UCAO-World Bank Development Project. Haut plateau de L'Ouest Technical Report No 9. Soil Science Department IRA-FONAREST ELOMA (Cameroon).

Faurie, C. (2003) *Ecologie. Approche scientifique et pratique*. Paris: Tec and Doc 5th edition, 407 p.

Finegan, B. (1992) The management potential of neotropical secondary lowland rain forest. *For. Ecol. Mgmt.* 47: 295-321.

Fotsing, J.M. (1992) Stratégies paysannes de gestion de terrains et de LAE en pays Ba-
    miléké Ouest Cameroun. *Bull. Réseau Erosion* 12: 241-254.

Fridley, J.D., Peet, R.K., Wentworth, T.R. & White, P.S. (2005) Connecting fine- and
    broad-scale species-area relationships of southeastern US flora. *Ecology* 86: 1172-
    1177.

Frontier, S. & Pichot-Viale, D. (1998) *Ecosystèmes. Structure, Fonctionnement, Evolu-
    tion.* Paris: Dunod, 445 p.

Givnish, T.J. (1999) On the causes of gradients in tropical tree diversity. *Journal of
    Ecology* 87: 193-210.

Gonga, X.Y., Brueck, H., Giese, K.M., Zhang, L., Sattelmacher, B. & Lin, S. (2008) Slope
    aspect has effects on productivity and species composition of hilly grassland in the
    Xilin River Basin, Inner Mongolia. *China. Journal of Arid Environments* 72: 483-
    493.

Gosselin, M. & Laroussinie, O. (coord.) (2004) *Biodiversité et gestion forestière.
    Connaître pour préserver. Synthèse bibliographique.* Paris: Cemagref Éditions,
    ECOFOR, Études, gestion des territoires 20, 320 p.

Gotelli, N., Colwell, R.K. (2001) Quantifying biodiversity: Procedures and pitfalls in
    the measurement and comparison of species richness. *Ecology Letters* 4: 379-391.

Harman, W.N. (1972) Benthic substrates: their effect on freshwater mollusca. *Ecology*
    53: 271-277.

Hooper, D.U., Chapin, F.S., Ewel, J.J., Hector, A., Inchausti, P. & Lawton, J.H. (2005) Ef-
    fects of biodiversity on ecosystem functioning, A consensus of current knowledge.
    Ecological Monographs 75(1): 3-35.

Huston, M.A. (1994) *Biological diversity.* Cambridge, UK: Cambridge University Press.

John, R., Dalling, J.W., Harms, K.E., Yavitt, J.B., Stallard, R.F., Mirabello, M., Hubbell,
    S.P., Valencia, R., Navarrete, H., Vallejo, M. & Foster, R.B. (2007) Soil nutrients
    influence spatial distributions of tropical tree species. *Proceeding of the National
    Academy of Science* 104: 864-869.

Jeeva, S., Kingston, C., Kiruba, S., Kannan, D. & Jasmine, T.S. (2007) Medicinal plants
    in the sacred forests of Southern Western Ghats. In: *National Conference on Recent
    Trends on Medicinal Plants Research* (NCRTMPR – 2007). Organized by Centre
    for Advanced Studies in Botany, University of Madras, Guindy Campus, Chennai
    – 600 025.

Kaczor, S.A. & Hartnett, D.C. (1990) Gopher tortoise (Gopherus polyphemus) effects
    on soils and vegetation ina Florida sandhill community. *American Midland Natu-
    ralist* 123: 100-111.

Khem, R.B. & Ole, R.V. (2003) Variation in plant species richness of different life forms
    along a Sub-tropical elevation gradient in the Himalayas, east Nepal. *Global Ecolo-
    gy & Biogeography* 12: 327-340.

Lamb, D., Parrota, J., Keenan, R. & Tucker, N. (1997) Rejoining habitat remnants; re-
    storing degraded Rainforest lands. In: Letouzey, R. (1968) *Étude Phytogéographique
    du Cameroun.* Paris: P. Le Chevalier.

Letouzey, R. (1985) *Notice de la carte phytogeographique du Cameroun*. Toulouse: Institute de la Carte Internationalede la Vegetation.

Libite, P.R. (2010) *La répartition spatiale de la population au Cameroun*. 6th ASSD-Cairo.

Lugo, A.E. (1992) Comparison of tropical tree plantations with secondary forest of similar age. *Ecol. Monographs* 62: 1-41.

Ma, M. (2005) Species richness versus evenness: independent relationship and different responses to edaphic factors. *Oikos* 111: 192-198.

Magurran, A.E. (1988) *Ecological diversity and its measurement*. Princeton, NJ: Princeton University Press.

Marini, L., Scotton, M., Klimek, S., Isselstein, J. & Pecile, A. (2007) Effects of local factors on plant species richness and composition of Alpine meadows. *Agriculture Ecosystems & Environment* 119: 281–288.

Mishra, B.P., Tripathi, O.P., Tripathi, R.S. & Pandey, H.N. (2004) Effect of anthropogenic disturbance on plant diversity and community structure of a sacred grove in Meghalaya, northeast India. *Biodiver Conser* 13: 421-436.

Mori, S.A., Boom, B.M., de Carvalho, A.M. & dos Santos, T.S. (1983) Southern Bahian moist forests. *Bot. Rev.* 49(2): 155-232.

Ohlemüller, R. & Wilson, J.B. (2000) Vascular plant species richness along latitudinal and altitudinal gradients: a contribution from New Zealand temperate forest. *Ecological letters* 3: 262-266.

Pélissier, P. (1980) *L'arbre en Afrique tropicale*. La fonction et le signe Cahiers ORS-TOM série Sciences Humaines XVII(3-4), 322 p.

Perry, D.A. (1994) *Forest Ecosystems*. Baltimore, MD: Johns Hopkins University Press, 649 p.

Peet, R.K., Wentworth, T.R. & White, P.S. (1998) A flexible, multipurpose method for recording vegetation composition and structure. *Castanea* 63: 262-274.

Pringle, C.M. (1990) Nutrient spatial heterogeneity: effects on community structure, physiognomy, and diversity of stream algae. *Ecology* 71: 905-920.

Reitalu, T, Sykes, M.T., Johansson, L.J., Lonn, M., Hall, K., Vandewalle, M. & Prentice, H.C. (2009) Small-scale plant species richness and evenness in semi-natural grasslands respond differently to habitat fragmentation. *Biological Conservation* 142: 899-908.

Rawat, S.D. (2011) Elevation reduction of plant species diversity in high altitudes of Garhwal, Himalaya, India. *Current science* 100(6): 833-836.

Rosenzweig, M.L. (1995) *Species diversity in space and time*. Cambridge, UK: Cambridge University Press.

Scarsbrook, M.R. & Townsend, C.R. (1993) Stream community structure in relation to spatial and temporal variation: a habitat template study of two contrasting New Zealand streams. *Freshwater Biology* 29: 395-410.

Schlosser, I.J. (1982) Fish community structure and function along two habitat gradients in a headwater stream. *Ecological Monographs* 52: 395-414.

Shannon, C.E., Weaver, W. (1949) *The mathematical theory communication*. Urbana II, University of Illinois Press.

Soomers, H., Derek K.D., Verhoeven, J.T.A., Verweij, P.A. & Wassen, M.J. (2012) The effect of habitat fragmentation and abiotic factors on fen plant occurrence. *Biodivers Conserv* 22(2): 405-424.

Sundarapandian, S.M., Chandrasekaran, S. & Swamy, P.S. (2005) Phenological behaviour of selected tree species in tropical forests at Kodayar in the Western Ghats, Tamil Nadu, India. *Cur. Sci.* 88: 805-809.

Swamy, P.S. Sundarapandian, S.M., Chandrasekar, P. & Chandrasekaran, S. (2000) Plant species diversity and tree population structure of a humid tropical forest in Tamil Nadu, India. *Biodiversity and Conservation* 9: 1643-1669.

Tonn, W.M. & Magnuson, J.J. (1982) Patterns in the species composition and richness of fish assemblages in northern Wisconsin lakes. *Ecology* 63: 1149-1166.

Swift, M.J. & Anderson, J.M. (1993) Biodiversity and ecosystem function in agro ecosystems. In: Schultze, E., Mooney, H.A. (eds) *Biodiversity and Ecosystem Function*. Spinger, NewYork, pp. 57-83.

Trumbore, S.E., Chadwick, O. & Amundson (1996) A rapid exchange between soil carbon and atmospheric carbon dioxide driven by temperature change. *Science* 272: 393-396.

UNEP-WCMC (United Nations Environment Programme – World Conservation Monitoring Centre) (2003) *WorldDatabase on Protected Areas (WDPA)*, Version 6. Compiled by the World Database for protected Areas Consortium, UK. August 2003.

Wilson, E.O. (2000) On the future of conservation biology. *Conservation Biology* 14: 1-4.

Vanpeene-Bruhier, S. (1998) *Transformation des paysages et dynamiques de la biodiversité végétale. Les écotones, un concept pour l'étude des végétations post-culturales. L'exemple de la commune d'Aussois (Savoie)*, PHD. Grenoble, ENGREF, T.1, 312 p., T.2, 127 p.

Venevsky, S. & Venevskaia, I. (2005) Hierarchical systematic conservation planning at the national level: Identifying national biodiversity hotspots using abiotic factors in Russia. *Biological Conservation* 12: 235-251.

White, F. (1983) *The Vegetation of Africa*. Paris: UNESCO.

Whittaker, R.H. (1975) *Communities and ecosystems*, 2nd edn. New York: MacMillan.

Zapfack, L., Engwald, S., Sonke, B., Achoundong, G. & Mandong, B.A. (2002) The impact of land conversion on plant biodiversity in the forest zone of Cameroon. *Biodiversity and Conservation* 11: 2047-2061.

Soil Quality Assessment of Cropping Systems in the Western Highlands of Cameroon

C.M. Tankou[1], G.R. de Snoo[2], G. Persoon[3] and H.H. de Iongh[2]*

Accepted for publication in the International Journal of Agricultural Research

[1] Faculty of Agronomy and Agricultural Sciences, University of Dschang, P.O. Box 222, Dschang, Cameroon.
[2] Institute of Environmental Sciences, Leiden University, P.O. Box 9518, 2300 RA Leiden, The Netherlands
[3] Department of Anthropology, Leiden University, P.O. Box 9518, 2300 RA Leiden, The Netherlands.
* Corresponding author; cmtankou@yahoo.com; tel: (237) 77 66 03 04; fax: (237) 33 45 15 66

Abstract

Soil nutrient depletion is a major constraint in agricultural development in sub-Saharan Africa. A study was conducted in the Western Highlands of Cameroon to assess soil nutrient balance and identify household and farm characteristics influencing soil nutrient balance using the procedures outlined in the NUTMON Tool-box. This is based on the assessment of the stocks and flows of nitrogen (N), phosphorus (P), and potassium (K) through the inputs (mineral fertilizers, organic inputs, atmospheric deposition and sedimentation) and outputs (harvested crop products, residues, leaching, denitrification and erosion). The gross margin was estimated as gross return minus variable costs while the net farm income was estimated as total gross margin minus fixed costs. The nutrient budgeting results revealed that nitrogen mining was very common at all levels with the greatest mining carried out by intercropping systems which generally received little or no off-farm inputs. High positive nutrient balances were found in market oriented crops. A general picture of the study site showed that only nitrogen was deficient while there were surplus amounts of potassium and phosphorus. The gross margins of green peppers, leeks and onions were negative while the others were positive. Legume intercrops could significantly modify the nutrient balance and sustainability in this region.

5.1   Introduction

The depletion of soil nutrients is a major constraint to sustainable agriculture (Smaling *et al.* 1993, 1996). Oenema *et al.* (2003) postulated that nutrient budgets of agroecosystems can be used as a tool to increase the understanding of nutrient cycling, or as a performance indicator and awareness raiser in nutrient management and environmental policy. The rural community of Cameroon in general and the Western Highlands of Cameroon (WHC) in particular, depend nearly entirely on agricultural activities to provide food, feed and income. The rising demographic pressure has imposed intensive land use over space and time which in turn demands high amounts of off-farm inputs. Though some studies have been done in the forest region of Cameroon (Stoorvogel & Smaling, 1990), no information so far exists in the WHC. Nutrient monitoring is a method that quantifies a system's nutrient inflows and outflows resulting in nutrient balance which is a useful indication of soil quality. Soil fertility is determined through the quantification of nutrient stocks and flows within the production systems (Bationo *et al.*, 1998; Deugd *et al.*, 1998). Nutrient balance can be determined at spatial scales ranging from field level to national level. A nutrient balance determined at the level of individual activities within a farm serves as a useful indicator, providing insight into the magnitude of the loss of nutrients from the system and the causes for such losses, which ultimately enables targeted intervention. Understanding the nutrient balance at each crop activity level within the farm and at farm level, can provide a useful guide when planning agricultural policy decisions to sustain the production system at these levels. The importance of this study to elucidate the nutrient dynamics in the breadbasket of Cameroon (Bergeret & Djoukeng, 1993) where no such study has been done before, cannot be overemphasized.

A nutrient balance is the difference between nutrient inputs and outflows (or losses). Positive balance for a particular nutrient means that the nutrient will accumulate in the soil while negative balance reflects the mining of the nutrient concerned. Accelerated rates of nutrient loss are evidence of soil depletion and are unsustainable over the long term. The accumulation of high levels of nutrients, particularly of P and N, are also undesirable, and are associated with increased pollutant export in the form of leachate or runoff (P and K), or in gaseous form through denitrification or volatilization (N). Nutrients are generally taken up by the plant from the solution phase of the soil, which is replenished through ion exchange, by dissolution from the solid mineral phase, or by mineralization of organic compounds. Some of the nutrients are usually returned to the soil in the form of crop residues and the remainder is removed from the field in the form of harvested products. Nutrients can also be lost through soil erosion, water runoff, soil sediment, leaching (mainly N and K) and volatilization (mainly N). The loss of nutrients is countered by biological

N-fixation, atmospheric deposition in rainfall, and the application of mineral fertilizers, animal manures, or compost.

Soil nutrient stocks are not static entities and studies in different parts of Africa at different spatial scales show that nutrients are being depleted at alarming rates (Stoovogel & Smaling, 1990; Van der Pol, 1992; Smaling et al., 1993; Smaling & Braun, 1996; Smaling et al., 1997; Scoones, 2001). Nutrients are annually taken away in crops or lost in processes such as leaching and erosion which far exceed the nutrient inputs through fertilisers, deposition and biological fixation (Smaling & Braun, 1996). Stoorvogel and Smaling (1990) estimated that 21 kg N, 2 kg P and 13 kg K were lost per ha and per year in southern Cameroon.

The concern for soil nutrient depletion and low soil fertility has led to the development of several integrated soil fertility management technologies that offer the potential for improving soil fertility management in Africa (Scoones & Toulmin, 1999). These include improved soil erosion control using living barriers or micro-catchments, inoculation of grain legumes for improved N-fixation, efficient use of manure and other locally available organic materials, use of green manure and cover crops (Delve & Jama, 2002) and use of low levels of N and P fertilisers on maize and beans (Wortmann et al., 1998; Wortmann & Kaizzi, 1998) in eastern Uganda. The application of nutrients to croplands is critically important for improving crop yields and the productivity of farmlands, however only approximately 50% of the applied nutrients are integrated into plant mass. The remaining nutrients accumulate in the soil, are emitted to the atmosphere (NO, NH_3, N_2O) or water bodies as soluble components (NO_3, PO_4) or as a component of soil. Agricultural activities are a significant contributor to the substantial increase of both nitrogen and phosphorus (Smil, 2000; Galloway, 1998) in our environment. These increases have contributed to the degradation of both air and water resources. A holistic approach to tackling the soil fertility problem through the integration of biophysical, socio-economic, institutional and policy factors has been proposed (CGIAR 2002). High input costs, inadequate knowledge of soil fertility management practices, the design of cropping systems exacerbated by pests and disease problems, global policies on the input-out market and institutional failures are amongst the determinants hampering the efforts of smallholder farmers (Van Reuler & Prins, 1993; Hart & Voster, 2006).

To address the importance of the problems faced by smallholder farmers of the WHC, the present research was undertaken by employing the NUT-MON-Toolbox to assess the nutrient balance at field (crop activity) level in the WHC. In addition, the study attempts to identify those household socio-economic and biophysical factors which influence soil quality and agro-economic performance and the profitability of cropping practices.

5.2 Materials and Methods

Site description: The study was carried out during the first and second cropping seasons in three villages: Bafou, Baleveng and Fongo-Tongo located in the WHC and found between the geographical coordinates of 05° 27' - 05° 37.62'N and 09° 57.502' - 10° 09.544' E (Figure 5.1). The general characteristics of the farm plot studied are given in Table 5.1. Soil samples were collected from the 17 farm plotsincluded in the study and analysed using standard procedures.

Table 5.1
Characteristics of farms studied

Characteristics	Average value (n = 17)	Standard deviation
Household size	7.18	6.11
Labour units*	2.12	1.36
Distance to the market (km)	2.91	1.52
PPUs	4.41	2.92
Area cultivated (ha)	0.17	0.19
Average slope (%)	11.58	7.92
Slope length (m)	53.61	32.29
pH-KCl	3.99	0.14
Total N (g/kg)	2.38	0.45
Extractable P (mg/kg)	27.08	3.6
Exchangeable K (mol/kg)	0.39	0.26
Organic C (g/kg)	6.03	1.29
Sand (%)	16.00	14.33
Silt (%)	48.00	19.71
Clay (%)	36.00	14.89

*Adult equivalence scale = people aged 18 years or over assigned a weight of 1; children below 18 years assigned a weight of 0.5.

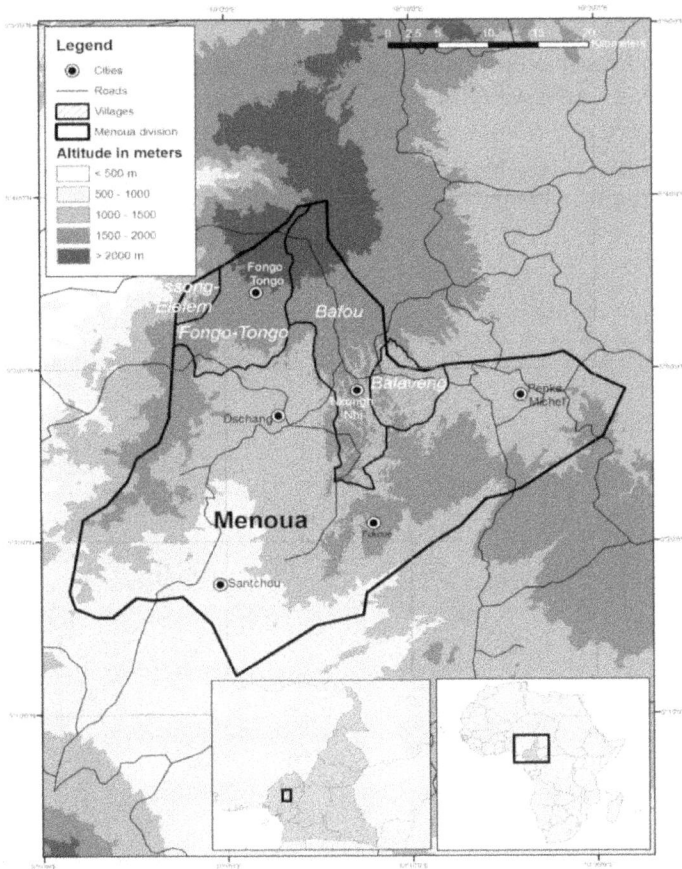

Figure 5.1
Geographical location of research site

5.2.1 Model description

NUTMON-Toolbox is software for monitoring nutrient flows and stock especially in tropical soils (Vlaming *et al.*, 2001). The toolbox enables the assessment of trends based on the local knowledge of soil fertility management and the calculation of nutrient balances. The tool is made up of a structured questionnaire, a database and two models (for calculating nutrient flows and economic parameters). The tool calculates the flows and balances of the macronutrients-N, P and K through an independent assessment of major inputs and outputs using the following equation.

Net soil nutrient balance = Σ (nutrient input) - Σ (nutrient output).

This is based on a set of five inflows (IN 1 to 5 are: mineral fertilizer, organic inputs, atmospheric deposition, biological nitrogen fixation and sedimentation), five outflows (OUT 1 to 5 are: farm products, other organic outputs, leaching, gaseous losses, erosion and human excreta) and six internal flows (consumption of external feeds, household waste, crop residues, grazing, animal manure and home consumption of farm products) as shown in Figure 5.2.

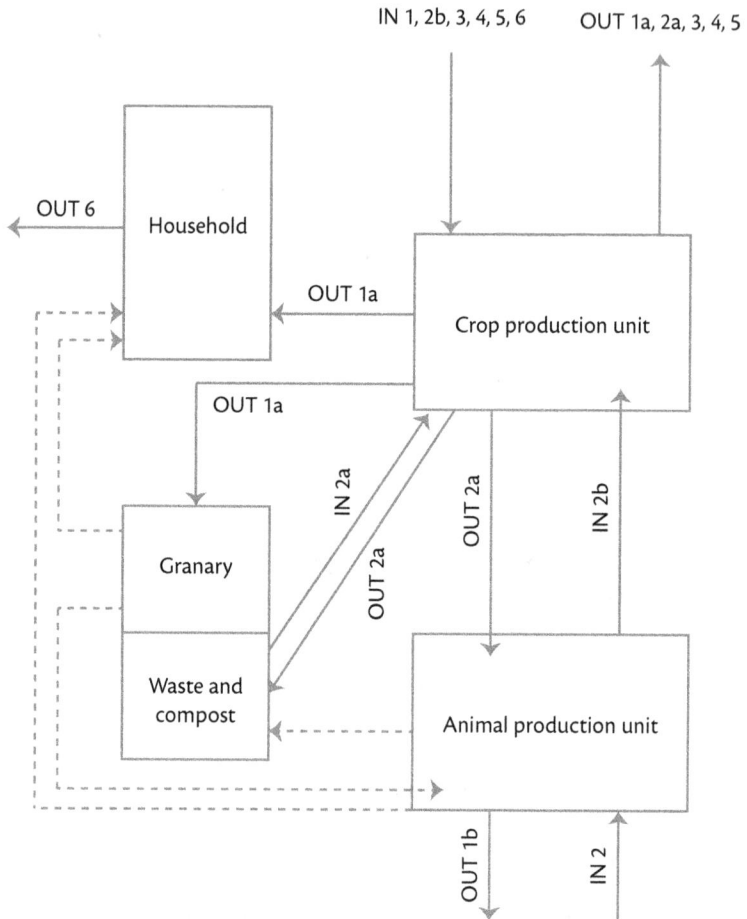

Figure 5.2
Nutrient flow within a farm in the Western Highlands of Cameroon

Farmers' interviews: The semi-structured questionnaire developed by Van den Bosche *et al.* (1998) was adapted to collect data from the heads of households of the selected farm plots. Biophysical, socio-economic and farming system data were collected through household interviews. Farmers gave information on the different production compartments, the different land uses and their major farm products and destinations. Nutrient flows directly related to their way of farming were quantified by questioning the farmers and through direct measurements on the farm or in the household. The inflows investigated by questioning the farmers were: the quantities of mineral fertilizers (IN 1), organic input such as manure (IN 2a) and organic fertilizers (IN 2b), entering the farm annually. The outflows included the quantity of crops (OUT 1a) and animal products (OUT 1b) leaving the farm as gifts or sales. Outflows measured were crop residue (OUT 2a), and animal manure (OUT 2b) leaving the farm. Farmers generally gave quantities in their own units, such as bundles, bags and buckets, which were converted to standard metric amounts. For each farmer, a field survey allowed us to identify the different land uses and the number of plots under each land use. Areas under different land use systems were measured using a hand-held GPS (Global Positioning System). This helped to estimate the different yields.

Use of transfer functions:
In this study the choice of the functions was based on studies conducted by Stoorvogel and Smaling (1990) on nutrient balances for sub-Saharan Africa. Wet atmospheric deposition was calculated from mean annual precipitation as follows:

Input of N (kg/ha/yr) = $0.14 \, p^{1/2}$
Input of P (kg/ha/yr) = $0.023 \, p^{1/2}$
Input of K (kg/ha/yr) = $0.092 p^{1/2}$, where p is the mean annual precipitation.

Biological nitrogen fixation (IN4) in production systems was estimated from the general equation:

IN4 (N) = [(legume field area × IN4a leguminous crop) + (area other crops × IN4b)] [total land area]$^{-1}$

IN4a is the symbiotically fixed and IN4b the non-symbiotically fixed nitrogen. It was assumed that 60% of the total N demand of leguminous crop (soybeans, groundnut and pulses) is supplied through symbiotic nitrogen fixation (Stoorvogel & Smaling, 1990).

IN4a (N) = [OUT1 (N) + Fl3a (N)] 60% + [(2 + P - 1350) × 0.005]

OUT1 (N) is the N exported in the leguminous crop, and FL3a, the quantity of N accumulated in crop residues.

The non-symbiotic nitrogen fixation was estimated from the function
IN4b (N) = 2 + (p - 1350) × 0.005 (Stoorvogel & Smaling, 1990)

Sedimentation (IN5) takes place in naturally flooded irrigated areas, where salinisation and iodisation naturally occur. No crop was registered on flooded or irrigated areas in the study site. This input was thus considered negligible. Deep capture (IN6) occurs in the presence of trees that exploit the soil layers below the normal root zone of crops. Intensive cultivation has eliminated such components in the system. This input was thus not considered in our study area.

Estimation of nutrient outflows not managed by farmers

Leaching below the root zone (OUT3)
In tropical soils phosphorus is tightly bound to soilparticles, and leaching involves only nitrogen and potassium. The quantity of N and K annually lost (kg ha^{-1} yr-1) was estimated from the transfer functions developed by Smaling *et al.* (1993).

For N leaching the Smaling 1993 model was used:

$OUT3 = (N_s + N_f) \times (0.021 \times p - 3.9)$ c < 35 percent

$OUT3 = (N_s + N_f) \times (0.014 \times p + 0.71)$ 35 percent < c < 55 percent

$OUT3 = (N_s + N_f) \times (0.0071 \times p + 5.4)$ c > 55 percent

Where:

$N_s =$ amount of mineralized N in the upper 20 cm of the soil;

       The mineralization rate of the site was estimated at 3% (Nye & Greenland, 1960).

$N_f =$ amount of N applied with mineral and organic fertilizers;

$p =$ annual precipitation (mm/year);

$c =$ Clay content of the topsoil (percent).

For K leaching, the Smaling 1993 model was used:

$OUT3 = (K_e + K_f) \times (0.00029 \times p + 0.41)$ c < 35 percent

$OUT3 = (K_e + K_f) \times (0.00029 \times p + 0.26)$ 35 percent < c < 55 percent

$OUT3 = (K_e + K_f) \times (0.00029 \times p + 0.11)$ c > 55 percent

Where:

$K_e =$ exchangeable K (cmol/kg);

$K_f =$ amount of K applied with mineral and organic fertilizers;

$p =$ annual precipitation (mm/year);

$c =$ clay content of the topsoil (percent).

OUT4 (gaseous losses) consists of two parts: gaseous N losses from the soil and gaseous N losses related to storage of organic inputs. Gaseous N losses from the soil are calculated as a function of the clay percentage and the precipitation:

$$OUT4 = (N_s + N_f) \times (-9.4 + 0.13 \times c + 0.01 \times p)$$

where:

$N_s =$ mineralized N in the rootable zone (kg/ha);

$N_f =$ N applied with mineral and organic fertilizer (kg/ha);

$c =$ clay content (percent);

$P =$ mean annual precipitation (mm/year).

Gaseous losses for N were calculated as:

$$N = Soil\ N + Fert\ N * (-9.4*0.13*clay\%+0.01p)$$

where

N = gaseous losses (kg/ha/yr)
Soil N = Mineralizable N in the upper 20 cm of the soil profile
Fert N = Mineral or organic fertilizer
Clay% = the clay content of the upper 20 cm of the soil profile
p = the mean annual precipitation.

Gaseous losses from animal dejections: (OUT 4b).
None of the farmers in the study area depended on animal production. As such animal dejections were not significant in the system. Gaseous losses from animal dejections were therefore assumed to be negligible.

OUT5 (erosion) was calculated using the USLE. A hypothetical soil loss per Farm Section Unit (FSU) was calculated based on slope, slope length, rainfall, soil characteristics and the presence of soil conservation measures. For each Primary Production Unit (PPU) or crop activity, the hypothetical soil loss (in kilograms per hectare per year) was multiplied by a crop cover factor, the nutrient content of the soil and an enrichment factor.

OUT 6 represents human faeces.

Nutrient balance
The NUTMON software quantifies nutrient flows in three ways: through the use of primary data, estimates and assumptions. Nutrient balances were quantified using the in-built transfer functions, equations and assumed values.

To distinguish between primary data and estimates, two different balances were calculated: the partial balance at farm level (IN1 + IN2) - (OUT1 + OUT2) made up solely of primary data and the full balance (ALL IN - ALL OUT) made up of a combination of the partial balance and the immissions (atmospheric deposition and nitrogen fixation) and emissions (leaching, gaseous losses, erosion losses and human excreta) from and to the environment.

The quantified economic flows reveal the profitability of farming activities (Vlaming *et al.*, 2001). Economic performance indicators were calculated at both activity level (crop) and farm household levels. The main indicators at activity level were gross margins (gross return minus variable costs) and net cash flows (cash receipts minus cash payments) per unit area. At farm household level, net farm income (total gross margin minus fixed costs) and family earnings (net farm income plus off-farm income) were the important indicators.
Correlation analysis, means, standard deviations and standard errors were calculated using the SPSS version 13.

5.3   Results

5.3.1 Description of the cropping system of sampled farmers based on interviews

The primary production units (PPUs) were sole cropping of vegetable crops practiced on fields generally between one and five kilometers from the homestead with the principal crops being potatoes (*Solanum tuberosum*), cabbages (*Brassica* sp), carrots *Daucus carota*), leeks (*Allium porrum*), onions (*Allium cepa*) and beetroots (*Beta vulgaris*), and intercropping around the homes. Multiple cropping was practiced on farms a few kilometers from the homestead with different combinations of maize (*Zea mays*), beans (*Phaseolus vulgaris*), potatoes (*Solanum tuberosum*), yams (*Dioscorea spp*), aroids (*Xanthosoma* sp and *Colocasia* sp) bananas and plantains (*Musa spp*). The intercropping systems identified from the sample of farmers were:
1 Beans + maize + potato (BMP),
2 Beans + maize (BM),
3 Beans + maize + potato + yam (BMPY) and
4 Maize + potato (MP).

Crop activities are predominant during the rainy season (mid-March to mid-November) and irrigation by gravity is used by some farmers during the off-season (mid-November to mid-March). Livestock or Secondary Production Unit (SPU) activities were uncommon with the farmers included in the nutrient balance study.

The identified nutrient flows into the farms in the research site were mineral fertilizersmostly 20.10.10 and urea (IN 1), off-farm chicken dung (IN 2), atmospheric deposition (IN 3), biological nitrogen fixation (IN 4a) and non-symbiotic nitrogen fixation (IN 4b). The only source of on-farm organic input were crop residue left after harvest which were directly recycled into the farm by incorporation during land preparation. A few farmers made compost around their homes used in home-gardens. In some cases the crop residues were burned and the ash exploited as organic input. Outflows from the farm included, crop uptake (OUT 1), removal in crop residue (OUT 2), gaseous loss (OUT 4), and erosion losses (OUT 5).

5.3.2 Nutrient balance at crop activity (PPU) level

The quantified nutrient balance at crop activity level (PPUs) using NUT-MON-Tooolbox are presented in Tables 5.2A, 5.2B and 5.2C for nitrogen, phosphorus and potassium respectively.

Table 5.2
NUTMON-Toolbox generated results.

A Average nutrient balance for Nitrogen for the different PPUs

| | Flows (kg) | | | | | | | | | Partial balance | Full balance |
| | Inputs | | | | Outputs | | | | | | |
	IN1	IN2	IN3	IN4	OUT1	OUT2	OUT3	OUT4	OUT5	(kg/ha/yr)	(kg/ha/yr)
Nitrogen											
PPU1 Beetroot (.04 ha)	1.25	1.18	0.11	0.05	-1.31	0.00	-0.83	-0.18	-1.24	11.44	-49.96
PPU2 Green pepper (0.29 ha)	51.50	22.44	0.95	0.37	-1.19	0.00	-91.14	-25.36	-19.42	254.49	-216.35
PPU3 Leek (0.21 ha)	11.00	15.11	0.41	0.21	0.00	-3.75	-9.35	-2.32	-1.83	208.11	151.95
PPU4 cabbage (0.09 ha)	13.33	19.50	0.13	0.12	-27.05	0.00	-22.39	-5.70	-3.39	132.29	-118.15
PPU5 carrot (0.08 ha)	10.56	6.01	0.27	0.15	-1.04	0.00	-27.29	-7.80	-6.66	1077.91	552.56
PPU6 Onion (0.01 ha)	20.00	8.73	0.64	0.32	-5.65	0.00	-49.99	-15.85	-9.37	115.25	-183.89
PPU7 Potato (0.19 ha)	27.21	6.31	0.48	0.31	-1.23	-0.03	-43.75	-13.14	-5.59	337.20	-96.25
PPU8 Tomato (0.14 ha)	77.63	4.73	0.09	0.06	-21.74	0.00	-33.35	-9.28	-0.57	939.65	382.54

	IN1	IN2	IN3	IN4	OUT1	OUT2	OUT3	OUT4	OUT5	Partial balance (kg/ha/yr)	Full balance (kg/ha/yr)
PPU 9 BMP (0.15 ha)	5.82	3.61	0.64	2.46	-2.51	-1.76	-40.49	-19.46	-94.13	49.44	-697.57
PPU10 BM (0.07 ha)	2.30	1.29	0.19	1.51	-3.50	-0.59	-15.44	-5.48	-11.82	22.58	-572.69
PPU 11 BMPY (0.12 ha)	5.10	3.89	1.12	2.51	-5.04	-1.14	-46.84	-30.60	-65.10	-6.36	-640.19
PPU12 MP (0.04 ha)	5.00	1.01	0.10	0.05	-0.17	-0.01	-14.14	-5.00	-7.39	144.87	-509.78

B Average nutrient balance for Phosphorus for the different PPUs

	Flows (kg)									Partial balance (kg/ha/yr)	Full balance (kg/ha/yr)
	IN1	IN2	IN3	IN4	OUT1	OUT2	OUT3	OUT4	OUT5		
Phosphorus											
PPU1 Beetroot (.04 ha)	0.28	0.31	0.02	0.00	-0.22	0.00	0.00	0.00	0.00	6.79	7.49
PPU2 Green pepper (0.29 ha)	4.40	5.87	0.16	0.00	-22.50	0.00	0.00	0.00	0.00	-42.77	-42.23
PPU3 Leek (0.21 ha)	1.65	3.95	0.07	0.00	0.00	-2.89	0.00	0.00	0.00	16.02	16.46
PPU4 cabbage (0.09 ha)	2.93	5.10	0.02	0.00	-2.12	0.00	0.00	0.00	0.00	89.88	90.10
PPU5 carrot (0.08 ha)	2.32	1.57	0.04	0.00	-0.43	0.00	0.00	0.00	0.00	256.54	256.87
PPU6 Onion (0.01 ha)	4.40	1.93	0.11	0.00	-1.09	0.00	0.00	0.00	0.00	25.70	26.21
PPU7 Potato (0.19 ha)	5.79	1.66	0.08	0.00	-0.37	0.00	0.00	0.00	0.00	75.18	75.63
PPU8 Tomato (0.14 ha)	16.50	1.24	0.01	0.00	-1.81	0.00	0.00	0.00	0.00	218.98	219.17
PPU 9 BMP (0.15 ha)	1.28	0.65	0.11	0.00	-0.41	-0.17	0.00	0.00	-0.01	12.06	12.63
PPU10 BM (0.07 ha)	0.23	0.21	0.03	0.00	-0.72	-0.09	0.00	0.00	0.00	-0.86	-0.36
PPU 11 BMPY (0.12 ha)	1.66	0.71	0.18	0.00	-1.02	-0.14	0.00	0.00	-0.01	0.78	1.70
PPU12 MP (0.04 ha)	1.10	0.27	0.02	0.00	-0.05	0.00	0.00	0.00	0.00	32.68	33.05

C Average nutrient balance for Potassium for the different PPUs

	Flows (kg)									Partial balance (kg/ha/yr)	Full balance (kg/ha/yr)
	IN1	IN2	IN3	IN4	OUT1	OUT2	OUT3	OUT4	OUT5		
Potassium											
PPU1 Beet-root (0.04 ha)	0.52	0.29	0.07	0.00	-1.46	0.00	-0.01	0.00	0.00	-41.63	-38.97
PPU2 Green pepper (0.29 ha)	8.99	5.59	0.62	0.00	-30.00	0.00	-0.12	0.00	0.00	-53.93	-52.20
PPU3 Leek (0.29 ha)	3.11	3.78	0.27	0.00	0.00	-0.56	-0.06	0.00	0.00	29.21	30.75
PPU4 cabbage (0.09 ha)	5.53	4.94	0.09	0.00	-22.85	0.00	-0.07	0.00	0.00	-138.68	-138.68
PPU5 carrot (0.08 ha)	4.38	1.50	0.18	0.00	-3.32	0.00	-0.04	0.00	0.00	292.00	290.61
PPU6 Onion (0.01 ha)	8.30	2.38	0.42	0.00	-7.25	0.00	-0.08	0.00	0.00	16.37	18.04
PPU7 Potato (0.17 ha)	10.92	2.05	0.32	0.00	-2.04	-0.02	-0.10	0.00	0.00	123.43	124.15
PPU8 Tomato (0.14 ha)	31.12	1.18	0.06	0.00	-14.50	0.00	-0.27	0.00	0.00	334.08	331.30
PPU 9 BMP (0.15 ha)	2.42	1.51	0.42	0.00	-1.23	-1.04	-0.02	0.00	-0.01	17.59	19.79
PPU10 BM (0.07 ha)	0.43	0.64	0.13	0.00	-1.27	-0.19	-0.01	0.00	0.00	6.00	8.02
PPU 11 BMPY (0.12 ha)	2.46	0.92	0.73	0.00	-2.03	-0.62	-0.02	0.00	-0.01	-2.54	1.16
PPU12 MP (0.04 ha)	2.07	0.50	0.06	0.00	-0.23	0.00	-0.02	0.00	0.00	58.18	59.27

The only positive full balances were observed with leek (PPU3), carrot (PPU 5) and tomato (PPU 8) for nitrogen. The highest negative full balance for nitrogen was observed with mixed intercropping of bean, maize and potato (PPU 9). All the intercropping systems showed high negative full balance for nitrogen. For phosphorus, the highest positive full balance was observed with carrot (PPU 5) and the lowest positive was observed with the beetroot (PPU 1). The highest negative full balance for phosphorus was observed with green pepper (PPU 2). For potassium, the highest positive full balance was observed with mixed intercropping of maize and potatoes (PPU 12) and the lowest positive full balance was observed with the intercropping of beans, maize, potato and yams (PPU 11). The highest negative full balance was observed with cabbage (PPU 4). With the partial NPK balances, except for the mixed intercropping system of beans, maize, potato and yams, all the others were positive for nitrogen while the partial balances for phosphorus and potassium followed trends similarto the respective full balances.

5.3.3 Nutrient balance for the research area:

With regard to the average results of the research site, the full balances were positive for potassium and phosphorus, which are less susceptible to loss and negative for nitrogen while the partial balances were positive for all three (Table 5.3).

Table 5.3

NUTMON-Toolbox generated average farm-level nutrient budget (kg/ha/yr) for the study site

Types of flow	N	P	K
Farmer managed			
IN1: Mineral fertilizers	126.25	24.95	47.07
IN2a: Organic fertilizer	57.71	14.82	16.57
IN2b: Grazing	0.00	0.00	0.00
OUT1: Farm products	-28.10	-9.32	-32.92
OUT2a: Farm residues	-4.62	-2.31	-1.34
OUT2b: Grazing	0.00	0.00	0.00
Partial budget 1	151.25	28.14	29.38
Not farmer managed			
IN3: Atmospheric deposition	2.91	0.48	1.91
IN4: Biological N fixation	4.60	0.00	0.00
OUT3: Leaching	210.57	0.00	-0.48
OUT4: Gaseous losses from the soil	-66.07	0.00	0.00
OUT5: Erosion	-65.83	-0.01	-0.01
Partial budget 2	-334.96	0.47	1.42
Total Budget	-183.71	28.61	30.80

Table 5.4 shows the yield and gross margin of the principal vegetable cash crops of the WHC.

Table 5.4
Yield and gross margin of the principal vegetable cash crops

Crop	Yield (kg/ha)	S.E.	Gross margin (US$/ha)[1]	S.E
Beetroot	4269.48	1412.34	42666.67	42666.67
Green pepper	988.91	663.34	-5617.82	6365.82
Leek	8364.75	3588.25	-3631.30	2749.11
Cabbage	7632.37	3016.41	1699.91	2046.87
Carrot	1992.52	584.49	1019.27	562.95
Onion	2274.28	605.24	-2910.39	1713.99
Potato	1319.09	204.64	69140.11	51709.27
Tomato	2619.87	1636.65	269.34	587.63

[1] 1 USD is approximately 500 FCFA

There were great variations in the yield and gross margin data and three of the crops cultivated showed negative gross margin values.

5.3.4 Household economics and farm characteristics

Table 5.5 shows the significant relationships between household economic and farm characteristics.

Table 5.5
Main significant correlations (Pearson) of household economic and farm characteristics

Characteristic	Positive correlation	Negative correlation
Farm N balance (kg/ha/yr)		Farm slope (%) (r = -0.67, r<0.01)
Farm P balance (kg/ha/yr)		Net farm income (US$) (r = -0.25, p<0.05)
Farm K balance (kg/ha/yr)	Total farm area (ha) (r = 0.55, p<0.05)	Farm slope (%) (r = -0.53, p<0.05)

There was a significant relationship between household economic character-istics and soil nutrient balance and also between farm characteristics and soil nutrient balance (Table 5.5).

5.4   Discussions

The nutrient balance results showed that adequate attention was given by the farmers to nutrient management of vegetable cash crops of the area. Similar results for market oriented crops were also noted by Surendran *et al* (2005). However De Jager *et al.* (1998) found a more negative N and K balance for higher market oriented crops in Kenya. Negative full balances for all three nutrients were seen with green beans and the bean+maize+potato+yam (BMPY) mixed intercropping system, implying that the amount of nutrients applied in these systems were sub-optimal. All three nutrients had positive balances with leek, carrot and tomato production. Such a situation needs to be checked because of the risk of environmental pollution, habitat destruction and the risk to human health and welfare (Hode, 1993).

As the yield-determining nutrient in most farming systems, adequate, but not excessive amounts of N are needed to sustain yields and contribute to the maintenance of soil organic matter (Goulding *et al.*, 2008).The net negative nitrogen balance for the research site could be attributed to the high outflow of N through harvested products, crop residues, losses from manure, leaching and gaseous loses as reported by Kroeze *et al.* (2003). Increasing the N fertilizer rates for crops grown, use of slow N release fertilizers or use of urease/nitrification inhibitors to improve Nitrogen Use Efficiency (NUE), growing insitu, incorporation of green manure during the fallow period to contain leaching losses that take place and the incorporation of organic manure to recycle nutrients in order to improve soil fertility, are the possible options to mitigate the negative N balance (Vos *et al.*, 2000). Nitrogen Use Efficiency canbe improved by applying the necessary nutrients in the correct amounts at the correct times (Krupnik *et al.*, 2004). The nitrogen uptake of the bean intercropping systems (BMP, BM and BMPY) was greatly influenced by the atmospheric nitrogen fixation (IN4) of the bean component. Traditionally, legumes have been viewed as excellent sources of N in agriculture (Kinzig & Socolow, 1994).

Nitrogen and potassium balances were found to be negatively influenced by the steepness of the farm slope similar to the findings of Schwab *et al.* (1993). Increasing population has resulted in farming of vulnerable areas like steep slopes. Practices like terraces and contour ploughing can help remedy the situation.

Full and partial balances of P were positive. This positive balance was mainly due to the optimal use of P fertilizers and the absence of pathways of losses of P other than crop uptake (OUT 1) and loss in crop residues (OUT 2). Use of P solubilizing (Phospho-bacteria) and mobilizing (VAM) microorganisms as biofertilizers could improve the utilization of native soil P in such situations of

P fertility buildup in soil (Debnath & Basak, 1986). The popular view that P is strongly held in soils has resulted in the build-up of excessive P levels in some soils, resulting in enhanced leaching (Heckrath *et al.*, 1995). Even where soil Plevels are at or near the optimum, the loss by erosion of small amounts of P adsorbed on sediments or insolution can trigger the eutrophication of freshwaters (Leinweber *et al.* ,2002). In this study, the household net farm income was negatively related to the soil phosphorus balance. This result is different from those of De Jager *et al.* (1998) who found that household net farm income showed no relation with the nutrient balance in three districts in Kenya.

Full and partial balances for K were also positive indicating an absence of deficiency in the study area. Similar results were found in the humid forest of Cameroon by Ehabe *et al.* (2010). Crop uptake of K is usually as much as N uptake and sometimes higher, as in the case of roots and tubers. There appear to be nohealth or environmental problems associated with K leaching and there are no gaseous emissions(Goulding *et al.*, 2008). However, over fertilization with potassium can induce a magnesium deficiency and also cause a loss of soil structure (Goulding & Annis, 1998). While the total farm area cultivated was positively correlated with the K balance in this study, De Jager *et al.* (1998) found in their study that the the cultivated area had a negative correlation with the level of nutrients leaving the farm through agricultural products sold.

The yields of the principal vegetable cash crops of the study area showed very high variations and the negative gross margins of green pepper, leek and onion indicated that either they were not profitable or that the cropping practices were not adequate. The negative gross margin of green pepper could be linked to the fact the cropping practices resulted in negative balances in all the three nutrients studied. Generally one of the causes of the low profit obtained by the farmers is due to widespread problems in the market and improving the relationship between farmers and buyers could contribute to a better economic situation for the farmers (Schipmann & Qaim, 2011).

5.5   Conclusion and recommendations

Resource flows and nutrient balances from this study show that soil nitrogen depletion is a major problem in the study area. Nutrient mining is more intense in the intercropping production system of the smallholder farmers in WHC but the inclusion of legume crops in the intercropping system alleviates the situation. Harvesting of crops for food and for sale and soil erosion are the most important sources of nutrient mining in the crop production systems. Therefore attempts to correct the imbalance need to address these and other socio-economic factors. Given the high costs of fertilizers, the recycling of kitchen residues, animal dejections and/or human faeces, and intensified use of legume as intercrops could significantly modify the nutrient balance and the sustainability of the systems. There exist findings that support the fact that intercropping enhances sustainable plant production (Ledgard, 2001; Aggarwal et al., 2002). There is good evidence that adding organic matter and fertilizers together improves NUE, as nutrients are held by the microbial biomass but that the microbial biomass plays animportant role in facilitating nutrient loss from soils insome situations (Turner & Haygarth, 2001). In research reported by Olesen et al. (2004), cover crops that reduced fertilizer N requirement by 27 kg/ha K1 and increased NUE from 42 to 52%, found however, that nitrate leaching increased by 14 kg/ha.

The magnitude of nutrient mining as a result of crop harvests in Africa is huge. Net losses of about 700 kg N. 100 kg P and 450 kg K per hectare during the past years have been estimated for 100 million ha of cultivated land which is the fundamental biophysical reason for the declining food production in smallholder farms in Africa (Sanchez et al., 1995; Smaling, 1993). Research findings have shown that leguminous trees in alley cropping systems can produce up to 20 t/ha/yr dry matter of prunings, containing as much as 358 kg N, 28 kg P, 232 kg K, 144 kg Ca, and 60 kg Mg (Young, 1989; Szottet al., 1991, Bowen, 1984), more than enough to meet most crop requirements. There exist many types indigenous soil enriching species such as *Tephrosia sp* and *Sesbania sp* including some recently introduced species in the WHC such as *Leucaena sp, Calliandra sp* and *Crotalaria sp* which if well managed, can significantly substitute chemical input in the system, prevent leaching, erosion and provide deep capturing of nutrients.

This study is one of the first attempts to create a database on nutrient balance studies in the WHC, providing a useful guide to policy makers, extension specialists and farmers alike in dealing with nutrient management.

Acknowledgements

This study was made possible by funding from Volkswagen foundation. The authors wish to thank Maarten van't Zelfde of the Institute for Environmental Sciences (CML), Leiden, for producing the map of the research site. We are very grateful to the extension workers of the Menoua Division who participated in the data collection and the household members who provided the useful data for the study.

References

Aggarwal, P.K., Garrity, D.P., Liboon, S.P. & Morris, R.A. (2002) Resource use andplant interaction in rice-mungbeanintercrop. *Agron. J.* 84: 71-78.

Baijukya, F.P. & De Steenhuijsen, P.B. (1998) Nutrient balances and theirconsequences in the banana based land use systems of Bukoba district, NorthWest Tanzania. *Agriculture, Ecosystems & Environment* 71: 147-158.

Bationo, A., Lompo, F. & Koala, S. (1998) Research on nutrient flows and balances in West Africa: state-of-the-art. *Agricult. Ecosyst. Environ.* 1: 19-35.

Bergeret, P. & Djoukeng, V. (1993) Evaluation économique des systèmes des cultures en pays Bamiléké (Ouest Cameroun). *Cahiers Agricultures* 2: 187-96.

Bowen, G.D. (1984) Tree roots and the use of soil nutrients. In: Bowen, G.D. & Nambiar, E.K.S. (eds) Nutrition of Plantation Forests, pp. 147-179. London, UK: Academic Press.

CGIAR (2002) Combating Soil Degradation in Africa. Consultative Group on International Agriculture Research Annual General meeting held in Washington DC, USA, October 30-31 2002. Document number AGM02/5/ii.

Debnath, N.C. & Basak, R.K. (1986) Effect of rock phosphate and basic slag on available P in acid soils in relation to soil characteristics, seasons and moisture regimes. *J. Ind. Soc. Soil Sci.* 34: 464-470.

De Jager, A., Kariuki, I., Matiri, F.M., Odendo, M. & Wanyama, J.M. (1998) Monitoring nutrient flows and economic performance in African farming systems (NUTMON). IV. Linking farm economic performance and nutrient balances in three districts in Kenya. *Agriculture, Ecosystems and Environment* 71: 81-92.

Delve, R.J. & Jama, B. (2002) Developing organic resource management options with farmers in eastern Uganda. 17th World Congress of Soil Science. Bangkok, Thailand.

Deugd, M., Roling, N. & Smaling, E.M.A. (1998) A new praxeology for integrated nutrient management, facilitating innovation with and by farmers. *Agricult. Ecosyst. Environ.* 71: 269-283.

Ehabe, E.E., Bidzanga, N.L., Mba, C.M., Njukeng, J.N., Inacio de Barros & Enjalric, F. (2010) Nutrient Flows in Perennial Crop-Based FarmingSystems in the Humid Forests of Cameroon. *American Journal of Plant Sciences* 1: 38-46.

Floret, C. (1998) *Raccourcissement du temps de jachère, biodiversité et développementdurable en Afrique Centrale (Cameroun) et en Afrique de l'Ouest (Mali, Sénégal).* Final Report, Comm. des Communautés européennes, Contrat TS3-CT93-0220 (DG 12 HSMU), IRD, Paris, 245 p.

Galloway, J.N. (1998) The global nitrogen cycle: changes and consequences. *Environ. Pollut.* 102: 15-24 Suppl. 1.

Gomez, A.A., Kelly, D.E., Syers, J.K. & Coughlan, K.J. (1996) Measuring Sustainability of Agricultural Systems at the Farm Level. Methods for Assessing Soil Quality. *SSSA Special Publication* 49: 401-409.

Goulding, K.W.T. & Annis, B. (1998) *Lime, liming and themanagement of soil acidity.* In Proc. no. 410, International Fertiliser Society, p. 36. York, UK: International Fertiliser Society.

Goulding, K., Jarvis J. & Whitmore, A. (2008) Optimizing nutrient management for farm systems. *Phil. Trans. R. Soc.* B 363: 667-680.

Hart, T. & Voster, I. (2006) *Indigenous knowledge on South African Landscape. Potential for African Development.* Cape Town, South Africa: Human Sciences Rsearch Council (HSRC) Press.

Heckrath, G., Brookes, P.C., Poulton, P.R. & Goulding, K.W.T. (1995) Phosphorus leaching from soils containingdifferent phosphorus concentrations in the Broadbalkexperiment. *J. Environ. Qual.* 24: 904-910.

Hodge, I. (1993) *Sustainability: Putting Principles into Practice. An Application to Agricultural Systems.* Paper presented to Rural Economy and Society Study Group, Royal Holloway College, December 1993.

Hoffman, I., Gerling, D., Kyogwom, U.B. & Mane-Bielfeldt (2001) Farmers management strategies to maintain soil fertility in a remote area in northwest Nigeria. *Agriculture, Ecosystems & Environment* 86(3): 263-275.

Howlett, D.J.B. (1996) Development of social, economic and biophysical indicators forsustainable land management in the South Pacific. In: Howlett, D. (ed.) *Sustainable Land Management in the South Pacific.* Network Document no. 19. International Board for Soil Research and Management. Bangkok, Thailand: IBSRAM.

Ikerd, J. (1993) Two related but distinctly different concepts: organic farming and sustainable agriculture. *Small Farm Today* 10(1): 30-31.

Juo, A.S.R. & Manu, A. (1996) Chemical dynamics in slash-and-burn agriculture. *Agriculture, Ecosystems & Environment* 58: 49-60.

Kinzig, A.P. & Socolow, R.H. (1994) Human impacts on the nitrogen cycle. *Phys. Today* 47(11): 24-31.

Kroeze, C., Aerts, R., Van Breemen, D., Van Dam, K. & Van der Hoek (2003) Uncertainties in the fate of nitrogen I. An overview of sources of uncertainity illustrated with a Dutch case research. *Nutrient Cycling in Agroecosystems* 66: 43-69.

Kotto-Same, J., Woomer, P.L. Moukam, A. & Zapfack, L. (1997). Carbon dynamics inslash-and-burn agriculture, and land-use alternatives of the humid forest zone in Cameroon. *Agriculture, Ecosystems & Environment* 65: 245-256.

Krupnik, T.J., Six, J., Ladha, J.K., Paine, M.J. & Van Kessel, C. (2004) An assessment of fertilizer nitrogen recoveryefficiency by grain crops. In: Mosier, A.R., Syers, J.K. & Freney, J.R. (eds) *Agriculture and the nitrogencycle. Assessing the impacts of fertilizer use on food productionand the environment.* SCOPE 65, ch. 14: 193-207. Washington, DC: Island Press.

Ledgard, S.F. (2001) Nitrogen cycling in lowinput legume-based agriculture, with emphasis on Legume/grass pastures. *Plantand Soil* 228: 43-59.

Leinweber, P., Turner, B.L. & Meissner, R. (2002) Phosphorus. In: Haygarth, P.M. & Jarvis, S.C. (eds) *Agriculture, hydrology and water quality*, ch. 2: 29-55. Wallingford, UK: CAB International.

Müller, S. (1996) *How to measure sustainability: an approach for agriculture and natural resources*. Discussion Paper series on sustainable agriculture and natural resources. IICA/BMZ/GTZ.

Müller, S. (1998) *Evaluating the sustainability of agriculture*. Eschborn, Germany: GTZ.

Oenema, O., Kros, H. & De Vries, W. (2003) Approaches and uncertainties in nutrient budgets: implications for nutrient management and environmental policies. *Eur. J. Agron.* 20: 3-16.

Olesen, J.E., Sørensen, P., Thomsen, I.K., Eriksen, J., Thomsen, A.G. & Berntsen, J. (2004) Integrated nitrogeninput systems in Denmark. In: Mosier, A.R., Syers, J.K. & Freney, J.R. (eds) *Agriculture and the nitrogencycle. Assessing the impacts of fertilizer use on food productionand the environment*. SCOPE 65, ch. 9: 129-140. Washington, DC: Island Press.

Palm, C.A. & Sanchez, P.A. (1991) Nitrogen release from the leaves of some tropical legumes asaffected by their lignin and polyphenolic contents. *Soil Biology & Biochemistry* 23: 83-88.

Rigby, D., Woodhouse, P., Burton, M. & Young, T. (2001) Constructing a Farm-LevelIndicator of Agricultural Sustainability. Sustainability. *Ecological Economics* 39: 463-478.

Salazari, A., Szott, L.T. & Palm, C.A. (1993) Crop-tree interactions in alley cropping systems onalluvial soils of the Upper Amazon Basin. *Agroforestry Systems* 22: 67-82.

Sanchez, P.A. (1976) *Properties and Management of Soils in the Tropics*. New York, NY, USA: John Wiley and Sons.

Sanchez, P.A. (1995) Science in agroforestry. *Agrofor. Syst.* 30: 5-55.

Schipmann, C. & Qaim, M. (2011) Supply chain differentiation, contract agriculture, and farmers marketing preferences: The case of sweet pepper in Thailand. *Food policy* 36(5): 667-677.

Scoones, I. (2001) *Dynamics and diversity: soil fertility and farming livelihoods in Africa: case studies from Ethiopia, Mali, and Zimbambwe*. London: Earthscan Publications Ltd., 244 p.

Scoones, I., Toulmin, C. (1999) *Policies for soil fertility management in Africa*. London: Department of International Development (DFID), 128 pp.

Szott, L.T., Fernandes, E.C.M. & Sanchez, P.A. (1991) Soil-plant interactions in agroforestry systems. *Forest Ecology & Management* 45: 127-152

Smaling, E. (1993) *An agro-ecological framework for integrating nutrient management, with special reference to Kenya*. Agricultural University of Wageningen, the Netherlands, 250 p. (Ph.D. Thesis).

Smaling, E.M.A., Fresco L.O. & de Jager, A. (1996) Classifying, monitoring and improving soil nutrient stocks and flows in Africa agriculture. *Ambio* 25(8): 492-496.

Smaling, E.M.A., Stoorvogel, J.J. & Windmeijer, P.N. (1993) Calculating soil nutrient balances in Africa at different scales. II District scale. *Fert. Res.* 35: 237-250.

Smaling, E.M.A., Stoorvogel, J.J. & de Jager, A. (2002) Decision making on integrated nutrient Management through the eyes of the scientist, the land-user and the policy maker. In: Vanlauwe, B., Diels, J., Sanginga, N. & Merckx, R. (eds) *Integrated plant nutrient management in sub-Saharan Africa: fromconcept to practice*, pp. 265-284. Wallingford, UK: CAB International.

Smaling, E.M.A., Braun, A.R. (1996) Soil fertility research in sub-Saharan Africa: New dimensions, new challenges. *Commun. Soil Sci. Plant Anal.* 27: 365-386.

Smaling, E.M.A., Nandwa, S.M., Janssen, B.H. (1997) Soil fertility in Africa is at stake. In: Buresh, R.J., Sanchez, P.A. & Calhoun, F. (eds) *Replenishing Soil Fertility in Africa*. Soil Science Society of America Special Publication No. 51. Madison, Wisconsin, USA: Soil Science Society of America, American Society of Agronomy, pp. 47-61.

Smil, V. (2000) Phosphorus in the environment: Natural flows and Human interferences. *Annu. Rev. Energy Environ.* 25: 53-88.

Stoorvogel, J.J. & Smaling E.M.A. (1990) *Assessment of soil nutrient depletion in sub-Saharan Africa, 1983-2000*. Report 28. Wageningen: The Winand Centre for Integrated Land, Soil and Water Research (SC-DLO).

Surendran, U., Murugappan, V., Bhaskaran, A. & Jagadeeswaran, R. (2005) Nutrient Budgeting Using NUTMON – Toolbox in an Irrigated Farm of Semi Arid Tropical Region in India – A Micro and Meso Level Modeling Study. *World Journal of Agricultural Sciences* 1: 89-97.

Schwab, G.O., Fangmeier, D.D., Elliot, W.J. & Frvert, R.K. (1993) *Soil and water conservation engineering*, 4th ed. New York, NY: Wiley, pp. 41-42.

Turner, B.L. & Haygarth, P.M. (2001) Phosphorus solubilisation in rewetted soils. *Nature* 411: 258-258.

Van der Pol, F. (1992) *Soil Mining. An Unseen Contributor to Farm Income in Southern Mali*. Bulletin of the Royal Tropical Institute No. 325, KIT Press, 47 p.

Vlaming, J.H., Van den Bosch, M.S., Van Wijk, A., De Jager, A.B. & Van Keulen, H. (2001) *Monitoring nutrient flows and economic performance in tropical forming systems (NUTMON)*. Publishers: Alterra, Green World Research and Agricultural Economics Research Institute, LEI, The Netherlands.

Van Reuler, H. & Prins, W.H. (1993) *The role of plant nutrients for sustainable food crop production in Sub-Saharan Africa*. Leidschendam, The Netherlands: Dutch Association of Fertilizer Producers (VKP), AK.

Vos, J. & Van Dder Putten, P.E.L. (2000) Nutrient cycling in a cropping system with potato, spring wheat, sugar beet, oats and nitrogen catch crops. I. Input and offtake of nitrogen, phosphorus and potassium Nutrient Cycling. *Agroecosystems* 56: 87-97.

Wortmann, C.S., Fischler, M., Alifugani, F. & Kaizzi, C.K. (1998) *Accomplishments of participatory research for systems improvement in Iganga District, Uganda 1993 to 1997*. OccasionalPublication Series, No. 27. Kampala, Uganda: CIAT, 40 p.

Wortmann, C.S. & Kaizzi, C.K. (1998) Nutrient balances and expected effects of alternativepractices in the farming systems of Uganda. *Agriculture, Ecosystems and Environment* 71: 117-131.

Young, A. (1989) *Agroforestry for Soil Conservation*. Wallingford, UK: International Council for Research and Agroforestry and CAB International.

6 General Discussion and Synthesis

The responses of the inhabitants of the Western Highlands of Cameroon (WHC) to an increase in population pressure and economic stress have brought about rural-rural mobility, rural-urban commuting, rural-urban migration, urban-rural migration and occupational diversification. The interaction between the human mobility patterns and farming systems, have effected change in the biodiversity and soil quality in the area. This study has made an attempt to bring into focus the interaction between the driving forces of human mobility, the sustainability of the farming systems and what has impacted on the biodiversity and soil quality in the WHC.

The demographic pressure of the WHC coupled with the cash crop crises prompted the rural population to seek for alternative ways of making a living. This brought about different types of human mobility.

6.1   What are the driving forces, contributions and categories of human mobility in the study area?

My results reveal that a diversification of occupation and income had become widespread in the WHC. One important reason for this was population growth and increased pressure on natural resources. In addition, the difficulties encountered by small-scale farmers in their struggle to make a living out of agriculture in an already risk-prone environment, had been exacerbated by economic reform and this had become an important 'push' factor for diversification. Young men in the WHC had found an occupational niche by offering transport services on motorbikes from the tarmac road and other pick-up points to the hinterlands. Return migration had injected financial resources and new skills into the local economy, but it was generally limited and varied between locations, depending on factors such as success in accumulating capital and skills, which in turn depended largely on educational levels and the income secured while away; the infrastructure and opportunities in home areas and access to local assets such as land, encouraged investment but also provided a safety net to migrants. In this my results support the views of Tacoli (2002) who stated that the contributions made by migrants when urban-rural migration takes place need to be better recognised and supported, as does their need for a safety net, given the often high levels of insecurity in urban labour markets. These migrants bring about significant diversification in the rural environment. Diversification is a key element in the livelihood strategies of most rural inhabitants and education is the most important single factor in determining whether diversification can start an accumulation process, or whether it is part of a survival strategy (Tacoli, 2002). Education would equally influence the understanding and adoption of technological packages.

Substantial information was obtained in order to provide adequate answers about the driving forces and their contributionto the different types of mobility found in the study area. While the low earnings from coffee production and demographic pressure were diagnosed as the most important driving forces on human mobility in the study area, various types of mobility and cropping systems were established as responses to alleviating problems and improvingreturns for the farmer. My research results show that there had been a transformationin the intra-rural mobility trends since the drop in the price of coffee on the international market in the early 90s. During the coffee boom-years, farm plots that were far from the homesteads were dedicated to subsistence crop production mostly through shifting cultivation or long fallow systems. However the end of the coffee boom and demographic pressure orchestrated rural-rural mobility for the production of vegetable crops that had become the replacement cash crops. As a result, farmers had started commuting to designated areas suitable for the cultivation of the vegetable cash crops and the shifting cultivation system subsequently gave way to intensive off-farm input systems. These results are in accordance with the observations of Potts (2006) who postulated that migration patterns can be a sensitive indicator of the spatial expression of economic and social variations in a region or a country. Increasing population and economic stress have acted as driving forces to reshape the lives of the farmers of the WHC. Results obtained from this work also demonstrated that better access to local assets (such as having access to farm plots at the high altitudes with a water supply) reduced the need for rural-urban migration, but non-economic aspects (such as household endogenous factors) were also significant determinants. Traditional approaches to migration focus on 'push' factors (economic hardship in areas of origin) and 'pull' factors (economic opportunities in destination areas) to explain the direction of movement (Lee, 1966). They usually fail to account for specific cultural and social factors, which play important roles in determining not onlydirection, but also composition and type of movement. Access to resources in home areas is a critical factor which is likely to be influenced by gender and age. In the WHC our findings showed that rural-to-urban migration was associated with both the number of household members and the age of the head of the household. The drop in coffee prices in combination with demographic pressure influenced the mobility pattern and land-use in the area.

In the study area, the commuting of rural farmers to the regional urban centres of Dschang where products were sold at higher prices to wealthier consumers was provoked by the costly off-farm inputs (improved planting materials, inorganic fertilizers and pesticides) that they were obliged to use due to the development of intensive cultivation practices; it was only in this way that the farmers could make their smallholding economically viable. The pull of urban markets for vegetable products that characterises the rural-urban commuting by high-input farmers of the WHC diagnosed by this study was also influenced by the level of the existing infrastructure. Given the perishabilityof these crops, long distance transportation without refrigeration facilities was impossible. The absence of appropriate storage and transportation facilities were major handicaps to evolution in theregion. The losses of perishable fruits and vegetables have been estimated at between 40 and 50% in the tropics and subtropics throughout the agro-food chain. This is mostly due to spoilage, physiological decay, and water loss, mechanical damage during harvesting, packaging and transporting (FAO, 1995 a and b). These results contribute to an understanding of intra-rural mobility in Africa which is currently the least studied aspect ofpopulation mobility (Potts, 2006) as opposed to international migration (Stanton et al., 1991; Crush and Wilmot, 1995; Crush and Mcdonald, 2002), both to neighbouring countries and the new interest in the role of the African diaspora far afield (Black & King, 2004). It is useful to note that rural-rural mobility has been identified as the strongest element of internal migration flows in some sub-Saharan African areas, with slightly more people involved than in the other three categories (rural-urban, urban-rural andurban-urban) put together (Potts, 2006). Rural-urban and especially urban-rural types of migration are much studied owing to their impacton urban economies (Potts, 1995, 2004, 2005; Ferguson, 1999; Beauchemin & Bocquier, 2004). Intra-rural movement is relatively neglected, perhaps because it is not associated with a shift in sectoral activity and is deemed to be of less interest in terms of assimilation issues (Potts, 2006). However, the massive movements of 'stranger' farmers within West Africa have been addressed, involving movements from the Sahelian zones to the wetter south to work oncash crops such as cocoa and coffee (Swindell, 1984), and some research has been done onthe return of refugees and internally displaced people to rural areas in their countries of origin (Wilson, 1992; Myers, 1994; Allen & Morsink, 1994; Koser, 1996, 1997). Host-stranger arrangements were widespread and flexible in both urban and rural areas and facilitated mobility in the WHC as in many other regions in Africa (Swindell & Iliya, 1999).

Efforts to maximize their efforts brought about achange in the farming systems with different approaches to the intensification of land use over space and time in addtion to the modification of input use in the farming systems. These new approaches modified the sustainability of farming systems.

6.2 What are the levels of sustainability and the relationship between sustainability and the different factors affecting farming systems in this zone?

During my research, I found that the traditional on-farm input-dependent system characterized by shifting cultivation and intercropping had fallen short of satisfying the demands of the rural population which was greatly concerned with income generation especially after the drastic drop in the market value of the original cash crop, coffee, in the early 90s. My study revealed a pragmatic and measurable format for sustainability relevant to small-scale farming in the WHC in particular, and sub-Saharan Africa in general. My results also revealed that the heavy dependence on off-farm inputs of the commercial vegetable producers of the WHC was a threat to the long-term sustainability of the system. The maintenance of agro-ecosystem function in the face of forces of change is a product of ecosystem properties and management activities. The main factors that influenced the long-term sustainability of the farming system in the WCH were grouped as: land use intensity over space, intensity of off-farm inputs, household adjustment factor and mobility of household, in descending order of importance. In other words, intensive land use was greatly favoured because of the possibility of cultivation during the off-season. In as much as this practice exploited the soil nutritive elements and destroyed the soil structure, it favoured the excessive use of off-farm inputs which were indispensable for the practice of intensive land use in my study area. These negative sustainability approaches in the WHC were found to be promoted by both younger and older farmers with less household residents. The last component in the multiple regression relation involved mobility of the household where favourable practices of crop biodiversity were discouraged by the distance of the farm plots from the homestead. The study showed that the further the farm plots were from the homestead, the greater was the tendency to have fewer crops in the field. The advantages of intercropping systems include yield stability under adverse environmental conditions, efficient use of limited growth resources, biological diversity, and potential control of pests and diseases.

My study revealed the litany of problems faced by the farmers in the WHC as laid out in this document. Smallholder farming systems are undergoing changes which impact on long term sustainability and social differentiation. In terms of livelihoods, many constraints inhibit both high quantity and quality outputs. Access to international markets for smallholder producers in the WHC was nonexistent, while local, national, and to some extent, regional marketing networks were the main outlet for small flows of diversified production. Investment in post-production infrastructure, information, roads and transport networks were found to be essential for the long term survival of these local marketing systems. High-quality land transport infrastructure facilitates economic activity, improves access to health care and education, raises social welfare by bringing consumer goods to rural areas, and facilitates social integration. In my research, I found that in the WHC, long distances, underinvestment in infrastructure, and inefficient trade and transport policies resulted in high land transport costs and low land transport quality. I concluded that this impeded the movement of labour and goods, reduced access to information and technology, and reduced export competitiveness by raising both the prices of imported inputs and the costs of transporting the final products to market, as was also noted by Farrington and Gill (2002). Continued access to farming assets is essential for poor and vulnerable groups in the WHC. Access to the main assets linked to farming (natural resources such as land and water, labour, either unpaid family or waged, credit, and markets) often depend on an individuals' social position (*ibid*). In my study, I noted that constraints over access to farming assets, especially land, were important factors in intergenerational conflict and changes in the pattern of employment. Increasingly, financial dependence on household heads, linked to contributing unpaid labour on the family's farm, was resented by the younger generations. Also Jambiya, (1998) and Mung'ong'o, (1998), found that shrinking natural resource availability and financial constraints on agricultural production strongly encouraged young men's engagement in non-farm employment as noted.

The modified farming systems as well as the extension of the land area for cultivation had greatly impacted the biodiversity of the agro-ecosystems in the area.

6.3 What are the different types of biodiversity of agro-ecosystems in the WHC and how are they influenced by abiotic factors?

The intensive cultivation system and drastic decrease in fallow length were other significant factors in the system which had already reduced plant biodiversity as reflected in the biodiversity of the 'sacred groves' of the area. For centuries, the indigenous people of the WHC and natives of other parts of Cameroon have loyally guarded patches of forest and accompanying streams commonly referred to as 'sacred groves' as described by Shonil and Claudia, (2006). They are believed to be the dwelling places of "Gods", and in some areas, are the burial grounds of royalty. I found that certain species of trees were considered sanctified (God's trees) and believed to have healing powers. The sacred quality of other groves derived from reverence for an animal species that lived there, most commonly monkeys or leopards. During my interviews I found that other sacredness originated from a river or stream that was home to a water god; the surrounding woods then became a protected area. Shonil and Claudia, (2006) found that logging, gathering of firewood, fires, hunting and cultivation in sacred groves were routinely banned, and in some groves, entry was restricted for women during their menstrual cycle. I also found that the custodians of the groves held ceremonies and rituals of ancestral worship in the forests. Telly (2006) noted that permission to take certain botanical elements for medicinal purposes from a sacred grove was sometimes granted.

I found that in the WHC, rural-rural mobility accounted for intensive landuse which had completely changed the structure of the native vegetation and caused severe plant diversity losses when compared with the composition of the sacred groves though some useful forage species such as *Pennisetum clandestinum* had been introduced to the area. I suggested that conservation of the sacred groves was important for the provision of certain ecosystem services in the WHC. Jackson et al (2007) already suggested that natural ecosystems provide many services essential to human existence such as water, medicinal plants and fruits. Increased species diversity provides more opportunity for species interactions, improved resource use, and ecosystem efficiency and productivity. It has been shown that intercropping in the grasslands outperforms the best monocultures, resulting in better production and more storage of carbon (Tilman *et al.*, 2002). In general, there is a positive correlation between species richness and productivity, and ecosystem resilience to drought (Tilman, 1997). Intercropping systems in the WHC were nearly exclusively in the production of subsistence crops. Well planned intercropping systems could provide better results in the cash crop systems of the area. Biodiversity in the WHC was strongly related to ecosystems and abiotic factors. I found that the herbaceous α-diversity[1] was significantly higher in the fallows than the sacred groves at low altitude; the tree species richness was higher at low altitude compared to the high altitude with tree ß-diversity[2] increasing with altitude; varying combinations of soil pH, total P, total K, CEC and slope percent were related to herbaceous species richness, herbaceous Shannon index and shrub species richness. In general the sacred groves contained a higher biodiversity, composed of trees, shrubs and herbaceous plants, compared to the fallow vegetation where trees and shrubs were very insignificant.

The different types of cropping and farming systems developed as a result of mobility and land use provoked significant changes in soil quality.

6.4 What is the impact of the modification of the farming system on soil quality at the crop and farm levels of the study area?

A modification of land-use intensity and cropping systems in the study area had a drastic impact on soil quality at the crop and farm levels. I found that nitrogen was the major element drained in the system while there was an excess of potassium and phosphorus in the predominantly sole cropping systems in the study area. This is in agreement with the findings of Matson *et al.* (1999) who noted that in tropical agro-ecosystems, most of the nitrogen is transformed into NO_3^- due to the heat and moisture and the majority is not taken up by plants but leached out of the system. My results show that the highest negative full balance for nitrogen wasto be found in mixed intercropping of bean, maize and potato; the highest negative full balance for phosphorus was in green pepper production;the highest negative full balance for potassium was observed in cabbage production.With the partial NPK balances, except for the mixed intercropping system of beans, maize, potato and yam, all the others were positive for nitrogen while the partial balances for phosphorus and potassium followed trends similar to the respective full balances. With regard to the average results of the research site, the full balances were positive for potassium and phosphorus, which are less susceptible to loss and negative

---

1 α diversity is the biological diversity at one site or sampling location.
2 β diversity is the change in diversity among various α diversities

for nitrogen while the partial balances were positive for all three. I also found that the yield and gross margins of the cropping systems in the WHC need some particular attention as negative gross margins were identified through my analysis. The negative gross margins for green peppers, leeks and onions indicated that either they were not profitable or the cropping practices were not adequate. Farmers may not be aware of this because the farm records they keep do not generally show such information.

6.5   Conclusion and recommendations

I found that vegetable cash crop farmers use significant amounts of inputs including a multitude of cultivars per crop which excludes them from the theory of involution or non-innovative intensification as outlined by Keys and McConnell (2005). I suggest that population pressure does not work in an unmediated fashion; rather it operates in conjunction with market and other social forces. The response of the inhabitants of the WHC with regard to the main forces that drive our framework can provide a guide when comparing their attitudes with respect to the many theories outlined. The famous models, of determinism (Malthus, 1989), possibilism (Boserup, 1965) as depicted by Gunnel (1997), and involution (Geertz, 1963) which still serve as a subtext to almost every contemporary debate on the environment and economics in the developing world, are cases in point; and yet, while the WHC cannot truly subscribe to environmental determinism (despite some isolated cases of detrimental effects on vulnerable lands) and involution, it does not conform entirely to the coordinates outlined by Boserup (1965). Non-innovative intensification or involution is not in essence, a typical characteristic of the WHC. The results obtained from my study show that demographic pressure does not lead to intensification of land use per se, as farm families employ strategies such as temporary or permanent migration and occupational diversification. This is substantiated by the significant numbers of household members involved in the different types of mobility in the WHC. By the same token, our results do not provide any justification for a Malthusian degradation spiral associated with increasing population in the WHC as suggested by Malthus (1989). Intensive agriculture in this zone is location and crop specific, a change which ocurredt after the precipitous decline in international coffee prices. Being nearly entirely dependent on agriculture, the practice is the farmer's way of coping with the economic situationin which they find themselves. However, no community is in perfect demographic equilibrium, and the number and ability of productive members, and their ratio to consumers, as well as the needs and aspirations of those consumers, are always important factors. Human movement, voluntary and involuntary, is a reflection of the initiatives and responses of people to the changing nature of society and the economy.

The different villages studied in this work clearly underline the necessity to tailor policies to local circumstances and to the specific needs and priorities of different groups in the region. It is crucial that future studies should account not only for internal demographic growth, but also for seasonal, generational, and permanent flows of labour and consumers, and to the knowledge, skills, and other resources carried with immigrants and returned migrants as proposed byKeys and McConnell (2005). No totally isolated agricultural communities can exist. Market signals are pervasive, influencing a broad range of outcomes. While proximate urban markets are seen to encourage the development of gardening and fruit production, distant international markets provide the incentive for planting other crops, particularly stimulants such as coffee, tea, and cocoa. These local or nearby markets encourage farmers to grow higher-value crops on limited landholdings (Eder, 1991). Farmers in the WHC are no exception to the rule as the national and international markets are the major stimulus guiding their farm activities. Apart from the local market in Dschang, the divisional headquarters, the Yaounde and Douala markets which are the political and economic capital cities respectively, in addition to Gabon and Central African Republic, international markets constitute the prime consumers of the farm products from the WHC. Eder (1991) also noted that farmers near Cebu City in the Philippines began cultivating new vegetables in response to increased demands and new taste preferences. A number of cases of horticulture also involved markets that lay farther away from cropping regions. Chilies, produced in southern Mexico, were sold some 1000 km away in Mexico City (Keys, 2004). The vagaries of the demand for stimulant crop products have important implications for the production of crops that take several years to produce fruit (Keys & McConnell, 2005). It would not be a surprise to observe a significant and abrupt change in the WHC if the market prices of coffee become very attractive again. Given the fact that coffee is the world's second most important commodity in legal internal trade after oil (O'Brienand Kinnaird, 2003) and that it is also the developing world's most important earner of capital, many farmers will be tempted to revert to it when the economic conditions become favourable. Improving the infrastructure and marketing channels is indispensable for sustainable production. The quality of land transport infrastructure is correlated with output, productivity, growth rates, land values, and market development (Farrington & Gill, 2002).

Sustainable agriculture is not a single, well-defined end goal. There is the general consensus that less time should be spent defining sustainable agriculture and more time spent on working to achieve it. This is because sustainability is a question rather than an answer or a direction rather than a destination. What constitutes sustainability in environmental, social, and economic terms is continuously evolving and is influenced by contemporary issues, perspectives, and values (Brodt et al., 2011). Agricultural systems are artificial systems and distinct from natural systems as a result of man's purposeful inter-ference and manipulation. The continued disturbance through artificial soil, weed and pest management exacerbated by the harvesting of desired plant and animal products influences the economic and ecological sustainability of the system. In the short term, economic sustainability may not be equal to ecolog-ical sustainability but in the long term this is expected to be the case. At the end of this study there will be a need to address the question of the economic and ecological sustainability of the WHC. How close to development are the activities carried out in the WHC that meet the needs of the present without compromising the ability of future generations to meet their own needs? The change from shifting cultivation to intensive land use has brought about mod-ifications, amongst which are biodiversity loss and improved livelihood from vegetable cash crop cultivation. This biodiversity loss is felt both with regard to cultivated crops and important natural vegetation which possess a number of positive ecological functions. Intensive land use around areas with water avail-able for irrigation also contributes to influence the sustainability of the system. It has been shown that natural resource scarcity is often an expression of con-flicting relationships between humankind and nature (Farber, 2000). Efforts are required to improve the sustainability of the traditional high biodiversity base system where the yield levels are presently very low. Environmental deg-radation undermines the economic and societal conditions essential to the well-being of a growing human population (Homer-Dixon, 1994; Scheffran, 1999). However, resource scarcity may not always lead directly to environ-mental conflicts, though a clear correlation may be hard to prove for complex situations. An increasing number of environmental agreements demonstrate that, under certain circumstances, threatened environmental systems can be improved through well designed approaches. In the case of the WHC, there is the possibility of exploiting natural resources that can improve both the soil quality and structure and promote less dependence on off-farm inputs. Appro-priate intercropping systems will also contribute to improving both the eco-nomic and ecological sustainability of the area. Expanding the cultivation of the natural vegetation found exclusively in the sacred groves and which is on the verge of extinction, is indispensable for a healthy functioning of the eco-system. In searching for ways to develop more sustainable agro-ecosystems, several researchers have suggested that tropical agro-ecosystems should mim-ic the structure and function of natural communities, (a practice followed by thousands of indigenous farmers for centuries), as these systems exhibit tight nutrient cycling, resistance to pest invasion, vertical structure and preserve bi-odiversity (Soule & Piper, 1992). If such an ecological approach is used, it will be important to ensure that promoted systems and technologies are suited to the specific environmental and socio-economic conditions of small farmers, without increasing risk or dependence on external inputs. I support the ide-as of Altieri (1995) who postulates that agro-ecological development projects should feature resource-conserving yet highly productive systems such as intercropping, agroforestry and the integration of crops and livestock. There is thus a great need to conserve the biodiversity of the WHC. Complementary genetic conservation through combining the use of insitu and exsitu technology (Damania, 1996) has been very successful in other areas. In situ conservation is vital because it provides a pathway for preserving complete biological diversity. It provides important basic necessities, such as medicines, fodder, food cosmetics, industrial products, fuel and timber, upon which humankind depends. Wild species, including relatives of cultivated plants, are crucial in cropimprovement programs as sources of genes for disease and pest resistance, environmental adaptability and nutritional qualities. In situ conservation evolves to establish and sustain a broad genetic base, stabilize and maintain populations and present opportunities for expanding agricultural systems (Chang, 1994). Great efforts are required in the WHC to preserve many species and cultivars of spices, food and vegetable crops that have become very rare since the introduction of new production techniques.

The capacity of the soil resource to perform the critical function of a life support system is undergoing unabated degradation of different kinds and deterioration due to nutrient depletion (Surendran et al., 2005). Keulen and Seligman (1992) showed that nutrient deficiencies limited the primary production more than water availability. Changes in soil fertility level should be monitored to provide an early caveat on adverse trends and to identify the problem areas. Understanding the nutrient balance at each crop activity level within the farm and at farm level is essential, in order to accurately guide agricultural policy decisions for planning at these levels, which can help to improve the crop yield and to sustain the production system. Tropical regions and especially the WHC experience this particularly as the NO_3^- is easily available in the dry season but is easily leached when the rains come at the beginning of the growing season before the young plant root systems arewell established. The introduction of nitrogen-rich leguminous species and other organic agriculture techniques can take care of this deficiency as was portrayed by intercropping systems that contained beans as a companion crop in the study area. However, a thorough understanding of farming systems is required in order to develop technological interventions, appropriate to the management of soil fertility (Hilhorst & Muchena, 2000). Chemical fertilizers could be regarded as the technical solution, directly as a supplier of nutrients, indirectly as a trigger for more nitrogen fixation and for more and better organic matter (manure, crop by-products and agroforestry). However, the efficiency of fertilizers is very low in African agriculture, in view of a low nutrient recovery of less than 30% (Van Duivenboden, 1992). Therefore, fertilizer-use must be combined with soil improvement. There is thus a need for research geared to evaluating the quantity, quality and synchrony of introducing different organic sources into the system.

Generally models remain important analytical tools as they organize the temporal evolution of very complex phenomena into different segments, each segment being composed of strikingly different characteristics, offering the means to centre attention on the most prominent aspects of reality. The proposed framework for this study (illustrated in Chapter 1) can be applied to both cross-section and longitudinal perspectives. The framework was applied in this study in a cross-section fashion. It has however, illustrated a clear relationship between the driving forces, responses and impacts in the agricultural production system in the Western Highlands of Cameroon. The scenarios portrayed in the conceptual framework were diagnosed by providing answers to:

- The driving forces and their contribution to the different categories of mobility
- The levels of sustainability, the relationship between sustainability and the different forces driving farming systems
- The different types of biodiversity and the influence of abiotic factors on agro-ecosystems
- The impact of the the modification of the farming system on soil quality at the crop and farm levels

In order for the population in the WHC to adjust to the drastic reduction of revenue from the cash crop of the area, trends in human mobility surfaced that in part transformed the traditional technology of shifting cultivation, to the intensive cultivation of novel species of cash crops based on high dependence on off-farm inputs. This accounted for the modification of soil quality at the crop and farm levels and the biodiversity of agro-ecosystems.

Future studies should explore the validity of the results of this study in other similar areas with pertinent historical data so as to expand and further develop the model presented here. However, the types of mobility claims advanced here are pertinent to the WHC; the quantitative evaluation of sustainability of farming systems reflects the reality on the ground; the constraints expressed by the farmers are contemporary; the nutrient flux evaluated at the crop and farm levels constitute a valuable database to guide future research.

The food challenge will be met using environmentally friendly and socially equitable technologies and methods, in a world with a shrinking arable land base, with less and more expensive petroleum, increasingly limited supplies of water and nutrients like nitrogen, and within a scenario of a rapidly changing climate, social unrest, and economic uncertainty (IAASTD, 2009). Across all farming systems, the efficiency of production or productivity measured in terms of output per unit input, linked to rewarding producers with fair prices and for a range of other environmental and social goods and services need to be emphasized. The only agricultural system that will be able to confront future challenges is one that will exhibit high levels of diversity such as are found in well managed multiple cropping systems, productivity with a positive energy flux, and efficiency in the use of production inputs. The positive indicators found in the study area included high crop biodiversity in most of the farms and the high use of organic inputs that mitigated the cost of production and increased the efficient use of production inputs. However, one of the key challenges in terms of food security is access to and distribution of the food produced.

References

Allen, T. and Morsink, H. (eds.) (1994) *When Refugees go Home: African Experiences.* James Currey: London.

Altieri, M.A. (1995) *Agroecology: the science of sustainable agriculture.* Boulder: Westview Press.

Beauchemin, C. & Bocquier, P. (2004) Migration and urbanizationin Francophone West Africa: a review of therecent empirical evidence. *Urban Studies* 41(11): 2245-2272.

Briggs, L. & Twomlow, S. (2002) Organic material flows within asmallholder highland farming system of South West Uganda. *Agriculture Ecosystems & Environment* 89: 191-212.

Black, R. & King, R. (2004) Transnational migration, returnand development in West Africa. Special Issue. *Population, Space and Place* 10(2): 75-174.

Boserup, E. (1965) *The Conditions of Agricultural Growth.* Chicago: Aldine.

Brodt, S., Six, J., Feenstra, G., Ingels, C. & Campbell, D. (2011) Sustainable Agriculture. *Nature Education Knowledge* 3(10): 1.

Chang, T.T. (1994) The biodiversity crisis in Asiacrop production and remedial measures. In: Peng, C.I. & Chou, C.H. (eds) *Biodiversityand terrestrial ecosystems.* Inst. Bot., Academia Sinica Monograph Ser. No. 14. Taipei: Academia Sinica.

Crush, J. & Wilmot, J. (eds) (1995) Crossing Boundaries: Mine Migrancy in a Democratic South Africa. Rondesbush: IDASA; Ottawa: IDRC.

Crush, J. & McDonald, D.A. (eds) (2002) *Transnationalismand New African Immigration to South Africa.* Montreal: Southern African Migration Project Queens/CAAS.

Damania, A.B. (1996) Field evaluation and utilization of collections of cereal genetic resources: The current status. *Indian J. Plant Genetic Resourc.* 9: 31-42.

Defoer, T., De Groote, H., Hilhorst, T., Kante, S. & Budelman, A. (1998) Participatory action research and quantitative analysisfor nutrient management in southern Mali. A fruitful marriage? *Agriculture Ecosystems & Environment* 71: 215-228.

Eder, J.F. (1991) Agricultural intensification and labor productivity in a Philippine vegetable gardening community: a longitudinal study. *Human Organization* 50 (3): 245-255.

FAO (1995a) Fruit and Vegetable Processing. *FAO Agricultural Services Bulletin* 119, Rome.

FAO (1995b) *Small Scale Post-harvest Handling Practices. A Manual for Horticulture Crops.* 3rd Edition, Series No. 8.

Farber, S. (2000) Welfare-based ecosystem management: an investigation of trade-offs. *Environ. Sci. Policy* 3: 491-498.

Farrington, J. & Gill, G. (2002) *Combining growth and social protection in weakly integrated rural areas.* Nat. Resource Perpect. No. 9. London: ODI.

Ferguson, J. (1999) *Expectations of Modernity: Myths and Meanings of Urban Life on the Zambian Copperbelt.* Berkeley, CA: University of California Press.

Geertz, C. (1963) *Agricultural Involution: The Process of Ecological Change in Indonesia.* Berkeley, CA: Univ. California Press.

Gunnel, Y. (1997) Comparative regional geography in India and West Africa: Soils, Landforms and Economic Theory in agricultural development strategies. *The Geographical Journal* 163(1): 38-46.

Hilhorst, T. & Muchena, F. (2000) *Nutrients on the move: soil fertility dynamics in African Farming systems.* London: International Institute forEnvironment and Development.

Homer-Dixon, T. (1994) Environmental scarcities and violentconflict: evidence from cases. *Int. Secur.* 19(1): 5-40.

IAASTD (International Assessment of Agricultural Knowledge, Scienceand Technology for Development) (2009) Agriculture at a Crossroads. In: *International Assessment of Agricultural Knowledge, Science and Technology for Development. Global Report.* Washington, DC: IslandPress.

Jackson, L.E., Pascual, U. & Hodgkin, T. (2007) Utilizing and conservingagrobiodiversity in agricultural landscapes. *Agric. Ecosyst. Environ.* 121: 196-210.

Jambiya, G. (1998) *The Dynamics of Population, Land Scarcity, Agriculture andNon-Agricultural activities: West Usambara Mountains, Lushoto District, Tanzania.* ASC Working Paper 28. Leiden: Africa Studiecentrum.

Keulen, H. van & Seligman, N.G. (1992) Moisture, nutrient availability and plant production in the semiaridregion. In: Alberda, Th., Keulen, H. van, Seligman, N.G. & Wit, C.T. de (eds) *Food from drylands. An integrated approach to planning of agricultural development.* Systems approach for sustainableagricultural development, 1. Dordrecht: Kluwer.

Keys, E. & McConnell, W.J. (2005) Global change and the intensification of agriculture in the tropics. *Global Environmental Change* 15: 320-337.

Koser, K. (1996) Information and refugee migration: the case of Mozambicans in Malawi. *Tijdschrift voor economische en sociale geografie* 87: 407-418.

Koser, K. (1997) Information and repatriation: the caseof Mozambican refugees in Malawi. *Journal of Refugee Studies* 10: 1-18.

Lee, E.S. (1966) A Theory of Migration. *Demography* 3: 47-57.

Malthus, T.R. (1989) *An Essay on the Principle of Population*, ed. P. James, 2 vols. Cambridge University Press for the Royal Economic Society.

Matson, P.A., McDowell, W.H., Townsend, A.R., & Vitousek, P.R. (1999) The globalisation of N deposition: ecosystem consequenses in tropical elements. *Biogeochemistry* 46: 67-83.

Myers, G. (1994) Competitive rights, competitive claims: land access in post-war Mozambique. *Journal of Southern African Studies* 20: 603-632.

Mong'ong'o, C. (1998) *Coming Full Circle: Agriculture, Non-Farm Activities and the Resurgence of Out-Migration in Njombe District*. ASC Working Paper 26, Leiden: Africa Studie Centrum

O'Brien, T.G. & Kinnaird, M.F. (2003) Caffeine and conservation. *Science* 300: 597.

Potts, D. (1995) Shall we go home? Increasing urban poverty in African cities and migration processes. *Geographical Journal* 161: 245-264.

Potts, D. (2004) Regional urbanization and urban livelihoods in the context of globalization. In: Potts, D. & Bowyer-Bower, T. (eds) *Eastern and Southern Africa: Development Challenges in a Volatile Region*. IBG/DARG Series. Harlow: Pearsons, pp. 328-368.

Potts, D. (2005) Counter-urbanization on the ZambianCopperbelt. Interpretations and Implications. *Urban Studies* 42: 583-609.

Potts, D. (2006). Rural Mobility as a Response to LandShortages: The Case of Malawi. *Population, Space & Place* 12: 291-311.

Scheffran, J. (1999) Environmental conflicts and sustainable development: a conflict model and its application in climate and energy policy. In: Carius, A. & Lietzmann, K.M. (eds) *Environmental Change and Securit*, pp. 195-218.

Shonil, A.B. & Claudia, R. (2006) Sacred groves:potential for biodiversity management. *Frontiers in Ecology and the Environment* 4: 519-524.

Soule, J.D. & Piper, J.K. (1992) Farming in nature's image. An ecological approach to agriculture. Washington DC: Island Press.

Stanton, R.S., Jacobsen, K. & Deane, S.W. (1991) *International Migration and Development in Sub-Saharan Africa*. Volume 1: Overview; Volume 2: Country Analyses. Discussion Papers, African Technical series 101 and 102. Washington, DC: World Bank.

Surendran, U., Murugappan, V., Bhaskaran, A. & Jagadeeswaran, R. (2005) Nutrient Budgeting Using NUTMON – Toolbox in an Irrigated Farm of Semi Arid Tropical Region in India – A Micro and Meso Level Modeling Study. *World Journal of Agricultural Sciences* 1(1): 89-97.

Swindell, K. (1984) Farmers, traders and labourers: dry season migration from northwest Nigeria. *Africa* 54: 3-18.

Swindell, K. & Iliya, M.A. (1999) Making a profit, makinga living: commercial food farming and urban hinterlands in North-West Nigeria. *Africa* 69: 386.

Tacoli, C. (2002) *Changing rural-urban interactions in sub-Saharan Africa and their impact on livelihoods: a summary*. Rural-urban working paper 7. London: IIED.

Telly, E.M. (2006) Sacred Groves, Rituals and Sustainable Community Development in Ghana. In: Schaaf, T. & Lee, C. (eds) *Conserving Cultural and Biological Diversity: The Role of Sacred Natural Sites and Cultural Landscapes.* Paris: UNESCO.

Tilman, D. (1997) Distinguishing betweenthe effects of species diversity and species-composition. *Oikos* 80: 185.

Tilman, D., Reich, P.B., Knops, J., Wedin, D., Mielke, T. & Lehman, C. (2002) Diversit-yand productivity in a long-term grasslandexperiment. *Science* 294: 843-845.

Van Duivenbooden, N. (1992) Sustainability in terms of nutrient elements with special reference to WestAfrica. Report 160. Wageningen: CABO-DLO, 261 p. + annexes.

Wilson, K. (1992) *Internally Displaced Refugees and Returnees from and in Mozambique.* Queen Elizabeth House, Refugee Studies Programme. Oxford: University of Oxford.

Summary

The Interactions of Human Mobility and Farming Systems and Impacts on Biodiversity and Soil Quality in the Western Highlands of Cameroon

Keywords

Human mobility, farming systems, sustainability, biodiversity, soil quality, Western Highlands of Cameroon.

This thesis draws on the findings from studies conducted between 2009 and 2011 in the Western Highlands of Cameroon. Population growth and a significant drop in the returns from the major cash crop for small farmers are the main drivers that have accounted for rural-urban and rural-rural interactions shaping the local economies and livelihoods of large numbers of people in Cameroon in general and the Western Highlands of Cameroon (WHC) in particular. The livelihood of the majority of the population, which is mostly poor, is being threatened by the rapid depletion of natural resources such as forests, and declining soil fertility. Principal threats to biodiversity in Africa include land use and land cover change, mainly through the conversion of natural ecosystems, particularly forests and grasslands, to agricultural land and urban areas. It is likely that land clearing and deforestation will continue and hence threaten genetic diversity as species loss occurs. The nature of farming is changing in many African countries because of demographic changes: the rural workers are migrating to urban areas, and many rural areas are becoming intensively cultivated. Trends in flows of people and information, and patterns of land use and occupational diversification reflect a dynamic process of economic, social and cultural transformation which needs to be better understood. The main objective of this research activity was to determine the interactions between farming systems and human mobility influenced by the drivers mentioned above, and the impact on local plant diversity and soil nutrient balance in order to develop guidelines to prevent land degradation and improve sustainability.

In order to better understand the determinants and impact of human mobility dynamics in the WHC, a comparative study was conducted through household and field surveys in three villages in the region and conceptualized based on the systems approach. The different types of mobility identified in this region were, rural-urban migration, urban-rural migration, rural-rural and rural-urban commuting as well as social mobility. A combination of household social factors (age of the head of the household and number of members in the household) significantly determined rural-urban migration. However, economic factors such as the possession of 'high-valued' farm plots discouraged this type of migration. 'High valued' farm plots were located at the high altitudes which provided the appropriate niche required for the production of vegetable cash crops. Rural-urban commuting was provoked by the need for conventional farmers to make ends meet. Commuting to the urban cities enabled them to sell their products produced under high cost off-farm inputs systems, at prices that could enable them make some profit. There was equally a high demand on land close to water bodies which favoured off-season production. The quest for 'high valued' farm plots and hired labour were the main determinants of rural-rural commuting. Urban-rural migration was orchestrated by a multitude of causes that rendered the urban milieu inappropriate. This type of migration contributed immensely to occupation diversification in the rural areas but also provoked social mobility.

Farming systems in the WHC have evolved over time, yielding both positive and negative contributions to rural welfare and livelihood. The traditional on-farm input-dependent system characterized by shifting cultivation and inter-cropping has fallen short of satisfying the new ambitions of the rural population which are highly concerned with income generation especially after the drastic drop in the market value of the original cash crop, coffee.In order to gain an insight into the farming system of the WHC, a field survey was carried out in three villages in this agro-ecological zone and analysed to understand the sustainability, general characteristics of the households and other drivers of the farming systems. The results revealed that the household characteristics were very similar across the villages while the sustainability though generally low, differed depending on the intensity of off-farm inputs in the production systems and other socio-economic factors. Sustainability had significant negative relationships with the intensity of land use, off-farm inputs, and sole cropping practice and a positive relationship with the age of the head of the household. The latent variables that influenced sustainability were land use intensity over space, intensity of off-farm inputs, the household adjustment factor and mobility of the household, in descending order of importance and they explained 62.15% of the total variation of sustainability in the study area.

Loss of biodiversity in the tropics is principally due to the destruction of habitat by anthropogenic activities especially the clearing of natural vegetation and conversion into agricultural cropland. Biodiversity provides many services essential to human existence. Increased species diversity provides more opportunity for species interactions, improved resource use, and ecosystem efficiency and productivity. In order to quantify the effects of land use on biodiversity in the WHC, an effort was made to investigate the extent of tree, shrub and herbaceous plant species richness in the sacred groves as dictated by topographic features and abiotic factors and quantify the impact of human disturbance through the evaluation of the herbaceous species in the fallowed lands with a view to generating baseline data of use to conservation. Data were collected at different altitudinal levels across the undisturbed (sacred groves) and disturbed vegetation (fallows) of the study area. The results revealed that sacred groves were rich in plant genetic diversity composed of a total of 42, 65 and 82 ethno-botanical species of herbs, shrubs and trees respectively, of varied ecological and economic importance. The herbaceous α-diversity was significantly higher in the fallows than the sacred groves at low altitude. The tree species richness was higher at low altitude compared to the high altitude with tree ß-diversity increasing with altitude. Varying combinations of soil pH, total P, total K, CEC and slope percent were related to herbaceous species richness, herbaceous Shannon index and shrub species richness. Intensive land-use has completely changed the structure of the native vegetation and caused severe plant diversity losses, though some useful forage species have been introduced in the area. Habitat changes in the sacred groves may be governed by biophysical drivers while a combination of human and biophysical drivers could be considered in the case of rotational fallow vegetation.

The rising demographic pressure has resulted in intensive land use over space and time which in turn demands high amounts of off-farm inputs. Studies were carried out in the WHC during the first and second growing seasons to evaluate the nutrient dynamics at the crop and farm levels. The nutrient budgeting results revealed that nitrogen mining was very common at all levels with the greatest mining carried out by intercropping systems which generally received little or no off-farm inputs. High nutrient budgets were found on market oriented crops. A general picture of the research site showed that only nitrogen was deficient while there were surplus amounts of potassium and phosphorus. The gross margins of green pepper, leeks and onions were negative. Legume intercrops could significantly modify the nutrient balance and sustainability in this region.

The determinants and types of mobility claims advanced here are pertinent to the WHC; the quantitative evaluation of sustainability of farming systems reflects the reality on the ground; the constraints expressed by the farmers are contemporary; the nutrient flux evaluated at the crop and farm levels constitute a valuable database. The magnificent contribution of this work should be indispensable to guide policy decisions and future research.

Samenvatting
De interacties tussen menselijke mobiliteit en land-bouwsystemen, en de impact op biodiversiteit en bodemkwaliteit in de Westelijke Hooglanden van Kameroen

Dit proefschrift is gebaseerd op de resultaten van onderzoek dat is uitgevoerd tussen 2009 en 2011 in de Westelijke Hooglanden van Kameroen. De belangrijkste oorzaken die hebben bijgedragen aan de interacties tussen de ruraal-urbane gebieden en de rurale gebieden onderling zijn bevolkingsgroei en een aanzienlijke terugval in de inkomsten van de belangrijkste *cash crop* voor kleine boeren. Deze wisselwerking heeft de lokale economieën en bestaansmogelijkheden van grote bevolkingsgroepen in Kameroen in het algemeen, en die van de Westelijke Hooglanden van Kameroen (WHC) in het bijzonder, gevormd. De bestaansmogelijkheid van de meerderheid van de arme bevolking wordt bedreigd door de snelle uitputting van natuurlijke hulpbronnen, zoals bossen, en de snel teruglopende bodemvruchtbaarheid. De belangrijkste bedreigingen voor biodiversiteit in Afrika omvatten de veranderingen in landgebruik en vegetatiebedekking, die worden veroorzaakt door de conversie van natuurlijke ecosystemen, in het bijzonder natuurlijke bossen en graslanden, in landbouwgrond en stedelijke gebieden. Het is waarschijnlijk dat deze ontginning en ontbossing zullen doorgaan. Daardoor zal de genetische diversiteit van soorten blijvend worden bedreigd. De aard van de landbouw verandert in veel Afrikaanse landen vanwege demografische veranderingen. Werkkrachten in de landbouw migreren naar de stedelijke gebieden en veel plattelandsgebieden worden intensiever gebruikt. Trends in de bewegingen van mensen en informatie tussen mensen, de patronen van landgebruik en trends van beroepsdiversificatie, weerspiegelen een dynamisch proces van economische en sociaal-culturele transformatie, die verder onderzocht moet worden. Het belangrijkste doel van dit onderzoek was het analyseren van de wisselwerking tussen landbouwsystemen en de menselijke mobiliteit, zoals die beïnvloed wordt door de bovengenoemde trends en factoren. Tevens is het onderzoek gericht op de impact van veranderend landgebruik en menselijke mobiliteit op de lokale diversiteit van planten en de balans van bodemnutriënten. Het uiteindelijke doel was om richtlijnen te ontwerpen ter voorkoming van landdegradatie en ter verbetering van de duurzaamheid van het landgebruik.

Om inzicht te genereren in de oorzaken en de impact van de dynamiek van menselijke mobiliteit in de Westelijke Hooglanden van Kameroen, is een vergelijkende studie uitgevoerd middels huishoud- en veld-onderzoek. Dit onderzoek is uitgevoerd in drie dorpen in de regio op basis van een experimentele onderzoeks-opzet. De verschillende mobiliteitstypen die, naast sociale mobiliteit, geïdentificeerd werden in deze regio waren: de ruraal-urbane migratie, de urbaan-rurale migratie, de ruraal-rurale migratie en het ruraal-urbaan forensisme. Een combinatie van kenmerken van sociale huishoudens (leeftijd van het hoofd van het huishouden en het aantal leden in een huishouden) droeg significant bij aan de ruraal-urbane migratie. Echter economische factoren, zoals het bezit van 'waardevolle' velden ontmoedigden dit type migratie. 'Waardevolle' velden waren gelegen op grotere hoogten, op terrassen die een geschikte niche vormden en die noodzakelijk waren voor de verbouw van dergelijke groentengewassen als *cash crop*. Ruraal-urbaan forensisme werd gestimuleerd door de behoefte van conventionele boeren om voldoende inkomen te verwerven. Door regelmatig naar de stad te trekken waren zij in staat hun producten, die met behulp van kostbare *off farm input* waren geproduceerd, te verkopen tegen prijzen die hen nog enige winst opleverden. Er was ook een sterke vraag naar land dichtbij waterbronnen die de productie mogelijk maken buiten het regenseizoen. Het zoeken naar 'waardevolle' velden en arbeid waren de belangrijkste determinanten van de ruraal-rurale migratie. Urbaan-rurale migratie werd veroorzaakt door een veelheid van factoren, waardoor het stedelijke milieu als ongeschikt werd gekwalificeerd. Dit type migratie droeg sterk bij aan de diversificatie van beroepen maar het heeft ook de sociale mobiliteit gestimuleerd.

De landbouwsystemen in de Westelijke Hooglanden van Kameroen zijn in de loop der jaren geëvolueerd en hebben zowel in positieve als negatieve zin bijgedragen aan de welvaart en de bestaansmogelijkheden op het platteland. Het traditionele systeem, met 'inputs' afkomstig van het eigen bedrijf, werd gekarakteriseerd door zwerflandbouw en gemengde teelt. Maar dit systeem schoot tekort in het bevredigen van de nieuwe ambities van de rurale bevolking die gericht zijn op het verwerven van nieuwe inkomsten, vooral na de drastische terugval in de marktwaarde van de originele *cash crop*, namelijk koffie. Om inzicht te verkrijgen in het landbouwsysteem van de Westelijke Hooglanden, werd een veldonderzoek uitgevoerd in drie dorpen in deze agro-ecologische zone. De gegevens van dit veldonderzoek werden geanalyseerd om zodoende de duurzaamheid van het landgebruik beter te begrijpen en de algemene karakteristieken van de huishoudens en de andere factoren van de landbouwsystemen vast te stellen. De resultaten toonden aan dat de karakteristieken van de huishoudens vrijwel gelijk waren in alle onderzochte dorpen. De duurzaamheid echter, hoewel in het algemeen laag, verschilde afhankelijk van de *off farm input* in de productiesystemen en van andere socio-economische factoren. Duurzaamheid had significant negatieve relaties met de intensiteit van landgebruik, *off farm inputs*, en de praktijk van verbouw van een enkele gewas. Duurzaamheid had een positieve relatie met de leeftijd van het hoofd van het huishouden. De latente variabelen die de duurzaamheid beïnvloedden waren, in afnemende mate van belangrijkheid 'de intensiteit van landgebruik over langere tijd', 'de intensiteit van *off farm inputs*' en 'de mobiliteit van het huishouden'. Deze variabelen verklaarden 62.15% van de totale variatie aan duurzaam landgebruik in het studiegebied.

Verlies van biologische diversiteit in de tropen is voornamelijk het gevolg van vernietiging van de habitat door antropogene activiteiten. De voornaamste oorzaak is het verwijderen van de natuurlijke vegetatie en conversie naar landbouwgrond. Biodiversiteit verschaft veel ecosysteemdiensten die essentieel zijn voor het menselijk bestaan. Grote soortendiversiteit verschaft meer mogelijkheden tot interacties tussen soorten. Een grote diversiteit zorgt ook voor een efficiëntere benutting van natuurlijke hulpbronnen en het verbetert de efficiëntie en productiviteit van het ecosysteem. Om de effecten van landgebruik op biodiversiteit in de Westelijke Hooglanden te kwantificeren, werd een poging gedaan om de mate van rijkdom aan diversiteit van bomen, struiken en kruiden te kwantificeren in de zgn. 'heilige bossen'. Deze biodiversiteit wordt bepaald door topografische kenmerken en a-biotische factoren. Tevens is getracht de impact te kwantificeren van de mate van verstoring door mensen, door te kijken naar de soorten kruiden op de braakliggende velden. Op basis van deze gegevens kan gekeken worden naar mogelijkheden van bescherming van de 'heilige bossen'. Deze data werden verzameld op verschillende hoogtes in zowel ongestoorde- als in verstoorde vegetatie (braak liggende velden) in het studiegebied. De resultaten toonden aan dat de heilige bossen rijk waren aan genetische diversiteit van planten met een totaal van tussen 42-82 soorten kruiden, struiken en bomen met een grote variatie aan ecologisch en economisch belang. De α-diversiteit was significant hoger in de braakliggende velden dan in de heilige bossen op lagere hoogteniveaus. De rijkdom aan boomsoorten was groter op lagere hoogte vergeleken met de grotere hoogte waarbij boom-β-diversiteit toenam met de hoogte. Verschillende combinaties van bodem pH, totale P, totale K, CEC en hellingspercentage waren gerelateerd aan de rijkdom aan kruidensoorten, Shannon diversiteitsindex van de kruiden, en de rijkdom aan struiken. Intensief landgebruik veranderde de structuur van de oorspronkelijke vegetatie volledig. Dit leidde tot grote verliezen aan plantendiversiteit, hoewel sommige nuttige soorten geïntroduceerd zijn in het gebied. Veranderingen van habitat in de heilige bossen werden vooral veroorzaakt door biofysische factoren terwijl een combinatie van menselijke en biofysische factoren belangrijk zijn bij de vegetatie op de braakliggende landbouwvelden.

De toegenomen demografische druk heeft in de loop der jaren, en over een steeds groter gebied, geleid tot intensiever landgebruik. Deze intensievere vorm van landgebruik was weer aanleiding tot de aanwending van grotere hoeveelheden *off farm inputs*. Veldonderzoek werd uitgevoerd in de Westelijke Hooglanden gedurende het eerste en tweede groeiseizoen om de dynamiek van de nutriënten te onderzoeken op het niveau van de gewassen en de velden. Uit de resultaten van de nutriëntenbalans bleek dat het uitputten van stikstof zeer algemeen was op alle niveaus. De grootste uitputting vond plaats bij landbouwsystemen, waarbij meerdere gewassen verbouwd werden en waarbij weinig of geen *off farm inputs* gebruikt werden. Een hogere nutriëntenbalans werd aangetroffen bij de verbouw van marktgewassen. Een algemeen kenmerk van de onderzoekslocatie was dat er een tekort aan stikstof was terwijl er een overschot aan kalium en fosfor was. Een negatieve nutriëntenbalans werd gevonden bij groene peper, prei en uien. Stikstofbindende gewassen zouden de nutriëntenbalans aanzienlijk kunnen verbeteren en daarmee ook de duurzaamheid van landgebruik in de regio.

De determinanten en de typen mobiliteit die in deze studie zijn beschreven zijn kenmerkend voor de Westelijke Hooglanden van Kameroen. De kwantitatieve evaluatie van de duurzaamheid van de landbouwsystemen weerspiegelt hetgeen in het veld wordt aangetroffen. De belemmeringen zoals die door de boeren verwoordt worden, hebben betrekking op de huidige situatie. De informatie over nutriëntenstromen, onderzocht op het niveau van zowel de gewassen als de velden vormen een waardevolle wetenschappelijke database . Op basis hiervan kunnen beleidsbeslissingen genomen worden en die kunnen ook de richting van toekomstig onderzoek helpen bepalen.

Résumé
Interactions entre la mobilité humaine et les systèmes de production et impacts sur la biodiversité et la qualité des sols dans les Hauts Plateaux de l'Ouest Cameroun

Mots clés

Mobilité humaine, systèmes de production, durabilité, qualité des sols, Hauts Plateaux de l'Ouest Cameroun

La présente thèse est basée sur les études menées entre 2009 et 2011 dans les hauts Plateaux de l'Ouest Cameroun. La croissance démographique et la baisse significative des revenus des producteurs agricoles issus du caféier, principale culture de rente, sont les principaux facteurs à l'origine des interactions entre les milieux ruraux et urbains d'une part et entre milieux ruraux d'autre part. Ces facteurs ont aussi façonné les économies locales et les modes de vie des populations rurales au Cameroun en général et des Hauts Plateaux de l'Ouest en particulier. Les moyens d'existence de la plupart des populations rurales pauvres sont menacés par l'épuisement rapide des ressources naturelles telles que la forêt et les sols. Ces populations sont alors contraintes à modifier leurs modes d'utilisation des terres. Cependant, les changements d'utilisation des terres et de la couverture végétale dues à la conversion des écosystèmes naturels en terres arables représentent une menace sérieuse pour la biodiversité. Au fur et à mesure que le défrichement et la déforestation vont continuer, la diversité génétique sera menacée suite à la disparition des espèces qu'ils engendrent. La production agricole elle-même est aussi entrain de se transformer à cause des changements démographiques : la main d'œuvre rurale migre abondamment vers les zones urbaines, tandis que beaucoup de milieux ruraux sont cultivés intensément. Les migrations des personnes et les flux d'information, la diversité des modes d'utilisation et d'occupation des terres reflètent un processus dynamique de transformation économique, sociale et culturelle qu'on a besoin de mieux comprendre. L'objectif principal de cette recherche était de déterminer d'une part les interactions entre les systèmes de production et la mobilité humaine telle qu'influencée par les facteurs sur-mentionnés, et d'autre part, leur impact sur la diversité des plantes et le statut nutritif des sols, afin de développer des indicateurs pour prévenir la dégradation des terres et améliorer la durabilité de leur utilisation.

Pour avoir un regard profond sur les déterminants et les impacts de la dynamique de la mobilité humaine dans les hauts plateaux de l'Ouest Cameroun, une étude comparative a été menée en utilisant les enquêtes des ménages et le diagnostic des champs dans trois villages selon l'approche système. En outre, pour quantifier l'effet de l'utilisation des terres sur la biodiversité dans les hauts plateaux de l'Ouest Cameroun, la richesse des espèces d'arbres, des arbustes et des plantes herbacées, a été déterminée en fonction des facteurs topographique et abiotiques, par un inventaire. L'impact de la perturbation humaine a aussi été déterminé à travers l'évaluation des espèces herbacées dans les jachères. Les données étaient collectées à différents niveaux d'altitude dans la végétation non-perturbées (forêts sacrées) et dans la végétation perturbée (jachères) au sein de la zone d'étude. Les observations dans les champs ont été faites pendant le premier et le deuxième cycle de production pour évaluer les dynamiques des éléments nutritifs au niveau des cultures et au niveau des parcelles cultivées.

Les différents types de mobilité caractérisant la mobilité sociale qui ont été identifiés dans cette région sont : la migration des personnes des milieux ruraux vers les milieux urbains, la migration des milieux urbains vers les milieux ruraux, les commutations entres zones rurales et, des milieux ruraux vers les milieux urbains. Une combinaison des facteurs sociaux des ménages (âge du chef de ménage et effectif du ménage) a significativement déterminé la migration des milieux ruraux vers les milieux urbains. Néanmoins les facteurs économiques tels que la possession des parcelles de haute valeur décourage ce type de migration. Ces dernières sont situées en haute altitude, dans les zones favorables aux cultures légumières de rente. L'accès aux marchés urbains leur permet de vendre leurs produits à des prix plus rémunérateurs. Il existe aussi une demande élevé pour les parcelles proches des cours d'eau propices à la production de contre saison. La recherche des parcelles de haute valeur agricole et la disponibilité de la main d'œuvre salariée sont les principaux déterminants de la commutation d'une zone rurale à l'autre. La migration des milieux urbains vers les zones rurales est provoquée par une multitude de causes qui rendent les milieux urbains inappropriés. Ce type de migration a beaucoup contribué à la diversification des occupations dans le milieu rural et à la mobilité sociale.

Les systèmes de production dans les hauts plateaux de l'Ouest Cameroun ont évolué dans le temps en apportant des contributions positives et négatives sur le bien être et les moyens d'existences de la population. Le système traditionnel dominé par la culture itinérante et l'association des cultures n'arrive plus à satisfaire les nouvelles ambitions de la population rurale de plus en plus intéressée par la création de sources alternatives de revenus surtout après la chute drastique du prix du café, ancien produit agricole de rente.

Les résultats des enquêtes ont montré que les caractéristiques des ménages sont semblables pour les différents villages tandis que la durabilité de l'exploitation est différente en fonction de l'intensité des intrants hors-champ utilisés dans les systèmes de production et d'autres facteurs socio-économiques. La durabilité de l'utilisation des terres a une relation significativement négative avec l'intensité d'utilisation des terres et la pratique de la culture pure ; par contre, elle a une relation significativement positive avec l'âge du chef de ménage. Les variables latentes influençant la durabilité sont : l'intensité de l'utilisation de la terre dans l'espace, l'intensité d'utilisations des intrants hors-champ, les facteurs d'ajustement du ménage et la mobilité du ménage. Ces facteurs expliquent 62,15% de la variabilité totale de la durabilité dans la zone d'étude.

Les résultats des inventaires révèlent que les forêts sacrées sont riches en diversité génétique. Au total, 42, 65 et 82 espèces ethnobotaniques d'herbes, d'arbustes et d'arbres ont été identifiés avec des valeurs écologiques et économiques variés, respectivement. L'α-diversité des herbes est significativement supérieur dans les jachères par rapport aux forêts sacrées. La richesse des espèces des arbres est supérieure dans les basses altitudes par rapport aux hautes altitudes avec une augmentation de la β-diversité avec l'altitude. Des combinaisons variées du pH du sol, du P total, du K total, de la CEC et du pourcentage de la pente sont liées à la richesse d'espèces herbacées, l'index de Shannon des herbacées et la richesse des arbustes. L'intensité d'utilisation des terres a complètement changé la structure de la végétation native et a causé une perte énorme de la diversité des plantes, malgré le fait qu'il y a eu introduction de quelques espèces fourragères utiles dans la zone d'étude. Les changements dans les forêts sacrées peuvent être expliqués par les facteurs biophysiques tandis que des combinaisons entre facteurs humains et biophysiques peuvent être à l'origine des changements dans la végétation des jachères.

La croissance de la pression démographique impose l'intensification de l'utilisation des terres dans l'espace et dans le temps, avec des quantités élevées d'intrants hors-champs. Les bilans des éléments nutritifs ont montré que l'épuisement de l'azote est fréquent à tous les niveaux avec une baisse marquée dans les associations qui ne bénéficient que de très peu ou pas du tout d'intrants hors-champ. Les bilans élevés sont constatés sur les cultures de rente. Une image générale des sites de recherche a montré que seul l'azote était déficient tandis que le potassium et le phosphore étaient en excès. Les marges brutes du poivron, du poireau et de l'oignon sont négatives. Leur association avec les légumineuses peut modifier significativement l'équilibre des éléments nutritifs et la durabilité de la production agricole dans cette région.

Les déterminants de la mobilité humaine et les principaux types de mobilités ainsi déterminés sont pertinents pour la zone dans les hauts plateaux de l'Ouest Cameroun ; l'évaluation quantitative de la durabilité des systèmes de production reflète la réalité en place ; les contraintes exprimées par les cultivateurs de la zone sont contemporaines ; les flux des éléments nutritifs déterminés au niveau des cultures et des parcelles cultivées constituent une bonne base de données. Le travail effectué apporte une contribution fondamentale pouvant être valorisée dans les politiques visant le développement agricole dans les Hauts Plateaux de l'Ouest Cameroun et par les recherches futures.

Acknowledgements

I would like to express my sincere gratititude to the Volkswagen foundation for providing the financial support for my PhD studies. My thanks are also directed to Geert de Snoo, Hans de Iongh and Gerard Persoon for accepting to supervise me and for the enormous efforts made despite their tight programs to assist me in realising this work. I wish to thank the administration of the University of Dschang in general and the Faculty of Agronomy and Agricultural Sciences in particular for granting me the time factor required to complete this work.

This work was carried out as part of the project on Human mobility, Networks and Institutions for the Management of Natural Resources in Africa. I wish to thank the project management board for their cooperation and assistance: Michael Bollig, Mirjam de Bruijn, Francis Nyamnjoh, Paul Hebinck, Tangie Fonchingong, Ulrike Wesch, Webster Wande, Martin Solich.

I very much benefited through interacting with the other students of the project: Emmanuel Hanai, Patience Mutopo, Florian Silangwa, Pinimidzai Sithole, Evelyne Teghomo, Anne-Christina Achterberg-Boness and Willem Odendaal

I am very grateful to the staff and graduate students of the Institute of Environmental Sciences (CML) and the African Studies Centre (ASC), of Leiden University for their different contributions. Particular thanks go to Ton Dietz, Maarten van't Zelfde, Jory Sjardijn, Esther Philips, Susanna van den Oever, Jose Brittijn and Maaike Westra.

I wish to thank Joseph Seh, Joseph Lumumba and Francis Ndi Wepngong for their friendship during my visits to the Netherlands.

I am very cognisant of the contributions of Dr. Loius V. Verchot, Annelies Bus and Dr. Nico de Ridder as I journeyed through thick and thin in quest for this goal.

Many thanks go to the farmers, extension workers, and students of FASA and FS of the University of Dschang who participated in the research in data collection. Special thanks go to Walter Tacham for his tremendous contribution in the biodiversity study.

Finally, I thank my wife, children and relatives for their support.

This work is dedicated to my parents (Tankou Alphonse and Ngifor Agnes) of blessed memory.

Curriculum Vitae

Christopher Mubeteneh Tankou, son of Tankou Alphonse and Ngifor Agnes, was born in Mankon, Bamenda, Cameroon on the 28th of February 1958. He attended Mankon Boys' School, Saint Bede's College Ashing Kom, and CCAST Bambili. He was admitted into the National Advanced School of Agronomy (ENSA), Cameroon, in 1980 and graduated with an "IngénieurAgronome" degree in 1985. He received his Master of Science at the University of Florida, USA, in December 1989 (his MSc thesis was nominated for the Award of Excellence for Graduate Research). He entered the PhD program at Leiden University in 2009. From 1985 to 1986, he taught in the Regional College of Agriculture Bambili. He was recruited to teach in the defunct University Centre of Dschang, now the University of Dschang, in 1986. He is a staff member of the Department of Crop Science in Faculty of Agronomy and Agricultural Sciences of the Unversity of Dschang, Cameroon. He has been serving as Head of Service for Diplomas, Programs and Research in the Faculty of Agronomy and Agricultural Sciences since 1999. He also served as guest lecturer in the Higher Teacher's Training College (ENS), Bambili from 1985 to 1987 and presently serves as guest lecturer in UDM, Bangante and UEC, Bandjoun. He is an official reviewer of the Journal of Agriculture and Biodiversity Research (JABR).

He is married to Auslia Nain Kweh and they have together one daughter (Sonia Ngifor) and four sons (Gilles Nohchawoh, Basil Talla, Conrad Sokoundjou and Colman Tamboh).

www.ingramcontent.com/pod-product-compliance
Lightning Source LLC
Chambersburg PA
CBHW080044280326
41935CB00014B/1778